INVESTIGATING RENEWABLE ENERGY and SUSTAINABILITY

UPPER PRIMARY

Published by Prim-Ed Publishing
www.prim-ed.com

INVESTIGATING RENEWABLE ENERGY AND SUSTAINABILITY

Published by Prim-Ed Publishing 2010

Copyright© R.I.C. Publications® 2008

ISBN 978-1-84654-219-0

PR–0323

This master may only be reproduced by the original purchaser for use with their class(es). The publisher prohibits the loaning or onselling of this master for the purposes of reproduction.

Copyright Notice

Blackline masters or copy masters are published and sold with a limited copyright. This copyright allows publishers to provide teachers and schools with a wide range of learning activities without copyright being breached. This limited copyright allows the purchaser to make sufficient copies for use within their own education institution. The copyright is not transferable, nor can it be onsold. Following these instructions is not essential but will ensure that you, as the purchaser, have evidence of legal ownership to the copyright if inspection occurs.

For your added protection in the case of copyright inspection, please complete the form below. Retain this form, the complete original document and the invoice or receipt as proof of purchase.

Name of Purchaser:

_____

Date of Purchase:

_____

Supplier:

_____

School Order# (if applicable):

_____

Signature of Purchaser:

_____

Internet websites

In some cases, websites or specific URLs may be recommended. While these are checked and rechecked at the time of publication, the publisher has no control over any subsequent changes which may be made to webpages. It is *strongly* recommended that the class teacher checks *all* URLs before allowing pupils to access them.

View all pages online   **Website:** www.prim-ed.com

Teachers notes

This book is divided into six sections.

1. What is energy?

In this section, pupils discover the basic principles of energy:
- *Energy is used to make things move.*
- *Energy is something that is stored, ready to use and can be used up.*
- *Energy can change from one form to another (energy transfer).*

Pupils also learn about the creation of the non-renewable energy sources, the fossil fuels, and the effect using them has on the environment.

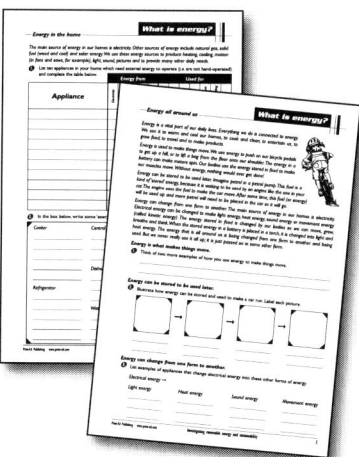

2. Solar energy   3. Wind energy   4. Hydropower

Each section includes a comprehensive informational text about the energy source, followed by a fun, hands-on investigation where pupils record their findings on the worksheets provided. The advantages and disadvantages of each energy source are also explored.

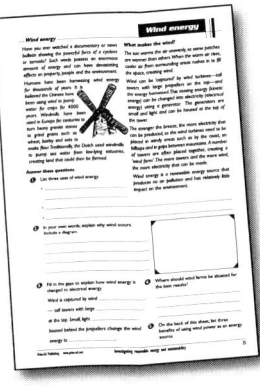 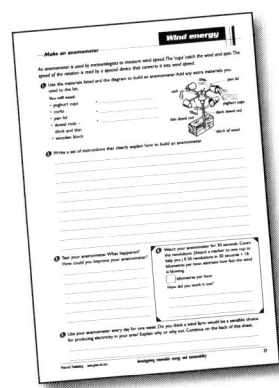 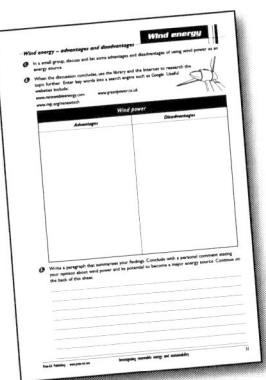

5. Alternative energy sources

In this section, pupils learn about four more renewable energy sources. These are:

- wave energy
- biomass energy
- tidal energy
- geothermal energy

Once again, an informational text and an investigation are provided for each energy source.

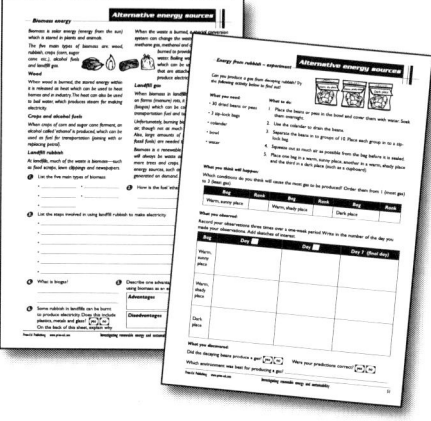

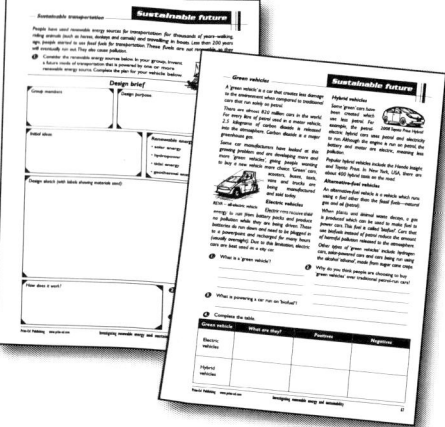

6. Sustainable future

In this section, pupils become aware of some ways to create communities and follow lifestyles which have a minimal impact on the environment. Pupils learn about sustainability while some different renewable energy sources are explored.

Activities to investigate energy-efficient homes and sustainable transportation, plus a group research project to determine energy sources in the community, are included.

Investigating renewable energy and sustainability

Foreword

Investigating renewable energy and sustainability explores those energy sources which have the potential to supply inexhaustible power without generating large volumes of pollution and waste, and which would allow us to reduce our dependence on fossil fuels.

Included are comprehensive overviews of the seven main types of sustainable energy being developed and/or used today:

- *solar energy*
- *wind energy*
- *tidal energy*
- *geothermal energy*
- *hydropower*
- *wave energy*
- *biomass energy*

Through contemporary comprehension activities followed by fun, hands-on investigations, pupils learn about the importance of these energy sources.

Investigating renewable energy and sustainability provides the frameworks for pupils to make discoveries and record their observations and ideas about renewable energy. The book also introduces the basic principles of energy, and offers strategies for sustainable living by looking at energy-efficient home design and 'green vehicles'.

The activities in *Investigating renewable energy and sustainability* cross major learning areas, with particular emphasis given to the key learning areas of *Science, Geography* and *English*.

Each activity is accompanied by a teachers page presenting background information, teachers notes, additional activities, display ideas and answers.

Investigating renewable energy and sustainability provides pupils with the opportunity to enhance their knowledge of the world around them and promote curiosity about planet earth and sustainable living.

Contents

Teachers notes	iv – v
Energy terms – Glossary	vi
Reflection sheet	vii
Assessment proforma	viii
Curriculum links	ix – xi

What is energy?
Energy all around us	2–3
Fossil fuels	4–5
Renewable energy	6–7
Energy in the home	8–9
Changing energy	10–11
Types of energy	12–13

Solar energy
The sun	14–15
Solar energy	16–17
Uses of solar energy	18–19
Build your own solar oven	20–21
Solar energy – advantages and disadvantages	22–23

Wind energy
Wind energy	24–25
Wind farms	26–27
Make an anemometer	28–29
Wind energy – advantages and disadvantages	30–31

Hydropower
Moving water	32–33
Hydroelectricity	34–35
Power of water investigation	36–37
Hydropower – advantages and disadvantages	38–39

Alternative energy sources
Wave energy	40–41
Energy from waves – experiment	42–43
Tidal energy	44–45
Tides and barrages	46–47
Biomass energy	48–49
Energy from rubbish – experiment	50–51
Geothermal energy	52–53
'Geothermal energy versus tourism' debate	54–55
Clueless crossword	56–57
Alternative energy sources – advantages and disadvantages	58–59

Sustainable future
What is sustainability?	60–61
Energy-efficient homes	62–63
Designing an energy-efficient home	64–6
Green vehicles	66–
Green vehicles – play	68
Sustainable transportation	7
Energy use in your community	

Teachers notes

Each pupil page has an accompanying teacher page, with information to assist the teacher with each lesson.

Background information has been included where necessary. For many activities, the factual information required for the lesson can be found on the pupil page.

In *Equipment/Materials*, the teacher is made aware of what needs to be collected prior to the lesson.

Answers for the activities on the worksheet are included. Some answers will need a teacher check; others will vary, depending on the pupil's personal experience or observations.

Teachers notes provide any additional information that may be necessary and include suggestions on possible ways to organise the lesson.

Publishing/Display ideas are suggestions for ways to present the resources used in the lesson or tasks completed by the pupils during the lesson.

Additional activities can be used to further develop the focus of the activity. They provide ideas to consolidate and clarify the concepts and skills taught.

Included in the front section of *Investigating renewable energy and sustainability* are:

- a *glossary* (page vi)

This is a short list of useful definitions relating to renewable energy. Space is provided for pupils to add their own words and definitions.

- a *reflection sheet* (page vii)

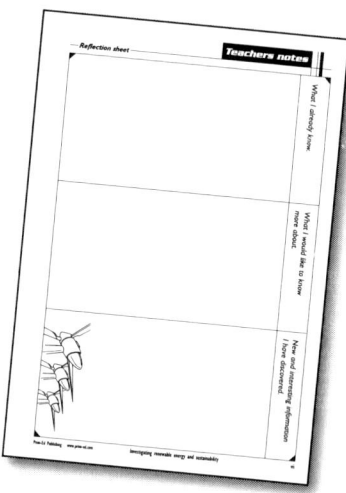

To be used at the beginning and end of the renewable energy unit, this forms a record of the pupil's learning and progress.

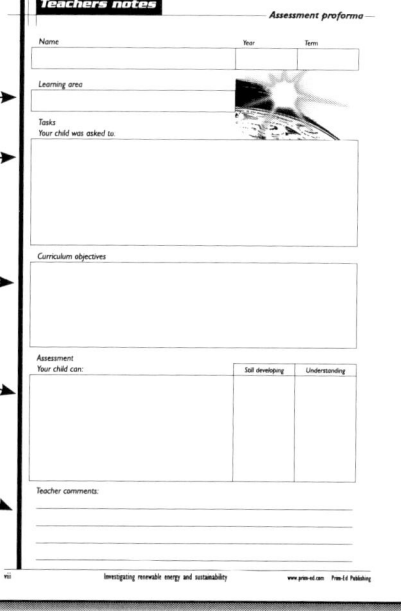

- an *assessment proforma* (page viii)

Fill in the appropriate learning area.

Give a brief description of the activity and what was expected of the pupils.

Write the relevant curriculum objectives.

Enter the curriculum objective(s) being assessed.

Use this space to comment on aspects of the individual pupil's performance which cannot be indicated in the formal assessment, such as work habits or particular needs or abilities.

- *curriculum links*
Found on page ix – xi.

Teachers notes

Energy terms – Glossary

biomass – organic matter such as plants and rubbish that can be used as an energy source

coal – a fossil fuel; formed by fossilised plants and burned in power plants to produce electricity

electricity – electric current used as a power source

energy – the ability (capacity) to do work

fossil fuel – naturally occurring fuel (e.g. coal, oil and natural gas), formed by the decomposition of dead plants and animals

fuel – something that is used to produce energy

geothermal – heat created from inside the Earth

hydroelectric – electricity produced from moving water

hydropower – energy that comes from the force of moving water

non-renewable – has a limited supply and cannot be replaced

pollution – materials harmful to living things

renewable energy – energy that is derived from the natural environment, such as sunshine, wind and water; can be replaced

solar energy – energy that comes from the sun

sustainable – able to supply our needs today as well as the needs of future generations

tidal energy – energy from the ebb and flow of the tide

wave energy – energy from wind blowing over the surface of the ocean

wind energy – energy that comes from the movement of air

My new energy terms and definitions

_____
_____
_____
_____
_____
_____
_____
_____
_____
_____
_____
_____
_____
_____
_____
_____

Reflection sheet

Teachers notes

| What I already know. |
| What I would like to know more about. |
| New and interesting information I have discovered. |

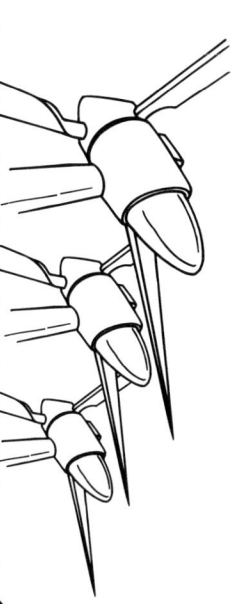

Teachers notes

Assessment proforma

Name

Year

Term

Learning area

Tasks
Your child was asked to:

Curriculum objectives

Assessment
Your child can:

	Still developing	Understanding

Teacher comments:

_____

_____

_____

_____

Curriculum links

Teachers notes

England – Key Stage Two	
Science	• know about ways the environment needs protection
Geography	• recognise how people can improve or damage the environment • recognise how decisions about environments affect the future quality of people's lives • recognise how and why people may seek to manage environments sustainably, and to identify opportunities for their own involvement
PSHE	• know that resources can be allocated in different ways and that these economic choices affect the sustainability of the environment
English	• obtain specific information through detailed reading • engage with challenging and demanding non-fiction subject matter
DT	• complete design and make assignments

Northern Ireland – Key Stage Two	
The world around us	• know ways people affect/conserve the environment, both locally and globally • know how we might act on a local or global issue • know about the consequences of change through investigating global issues • design and make models
Language and literacy	• use a variety of reading skills for different reading purposes

Teachers notes

Curriculum links

Republic of Ireland – 5th and 6th Classes	
Science	• design and make objects • become aware of the Earth's renewable and non-renewable resources • appreciate the ways in which people use the Earth's resources • appreciate the need to conserve resources • participate in activities that contribute to the enhancement of the environment • discuss local, national and global environmental issues • appreciate individual, community and national responsibility for environmental care
Geography	• recognise and investigate aspects of human activities which may have positive or adverse effects on environments • become aware of the Earth's renewable and non-renewable resources • appreciate the ways in which people use the Earth's resources • appreciate the need to conserve resources • discuss local, national and global environmental issues • appreciate individual, community and national responsibility for environmental care
SPHE	• appreciate the environment and develop a sense of responsibility for caring for the environment and being custodians of the Earth for future generations
English	• use comprehension skills and explore non-fiction texts

Curriculum links

Scotland – Second Level

Science	• research a major environmental or sustainability issue of national or global importance
Social studies	• discuss the environmental impact of human activity and suggest ways in which we can live in a more environmentally responsible way
Literacy and English	• identify and consider the purpose and main ideas of a text and show understanding by responding to questions
Technologies	• analyse how own lifestyle can impact on the environment and Earth's resources, and make suggestions about how to live in a more sustainable way • investigate the use and development of renewable and sustainable energy to gain an awareness of their growing importance in Scotland and beyond • meet design challenges and construct models

Wales – Key Stage Two

Science	• know how humans affect the local environment
Geography	• care for the environment and know the importance of being a global citizen • investigate 'geography in the news' in the local area and the wider world • ask questions about an environment or geographical issue • express their own opinions and be aware that people have different points of view about environments and geographical issues • consider how people have affected the environment and how it can be looked after
PSE	• know how the environment can be affected by the decisions we make individually and collectively
English	• read for different purposes • read extracts and texts with challenging subject matter that broaden perspectives and extend thinking
DT	• design and make products

What is energy?

Energy all around us

Objective:
- Reads texts and answers questions about energy, stored energy and energy transfers.

Teachers notes:
- Introduce the concept of energy. Ask pupils to share their impression of what energy is. Record their ideas.
- Pupils read the text at the top of the page then discuss it with a partner before completing the answers.

Publishing/Display ideas:
- Pupils design a large chart showing appliances (cut out or drawn) and label the energy source they use. Display the charts.

Answers:
1.– 3.   Teacher check.

Additional activities:
- Which foods give our bodies the most energy? How is this energy measured? (kilojoules – kJ) Look on food packaging to see how the amount of energy is displayed.
- Where does petrol come from? Pupils find out about the process of recovering crude oil to create fuels.

What is energy?

Energy all around us

Energy is a vital part of our daily lives. Everything we do is connected to energy. We use it to warm and cool our homes, to cook and clean, to entertain us, to grow food, to travel and to make products.

Energy is used to make things move. We use energy to push on our bicycle pedals to get up a hill, or to lift a bag from the floor onto our shoulder. The energy in a battery can make motors spin. Our bodies use the energy stored in food to make our muscles move. Without energy, nothing would ever get done!

Energy can be stored to be used later. Imagine petrol in a petrol pump. This fuel is a kind of 'stored' energy, because it is waiting to be used by an engine like the one in your car. The engine uses the fuel to make the car move. After some time, this fuel (or energy) will be used up and more petrol will need to be placed in the car so it will go.

Energy can change from one form to another. The main source of energy in our homes is electricity. Electrical energy can be changed to make light energy, heat energy, sound energy or movement energy (called kinetic energy). The energy stored in food is changed by our bodies so we can move, grow, breathe and think. When the stored energy in a battery is placed in a torch, it is changed into light and heat energy. The energy that is all around us is being changed from one form to another and being used. But we never really use it all up; it is just passed on in some other form.

Energy is what makes things move.

❶ Think of two more examples of how you use energy to make things move.

_____

_____

Energy can be stored to be used later.

❷ Illustrate how energy can be stored and used to make a car run. Label each picture.

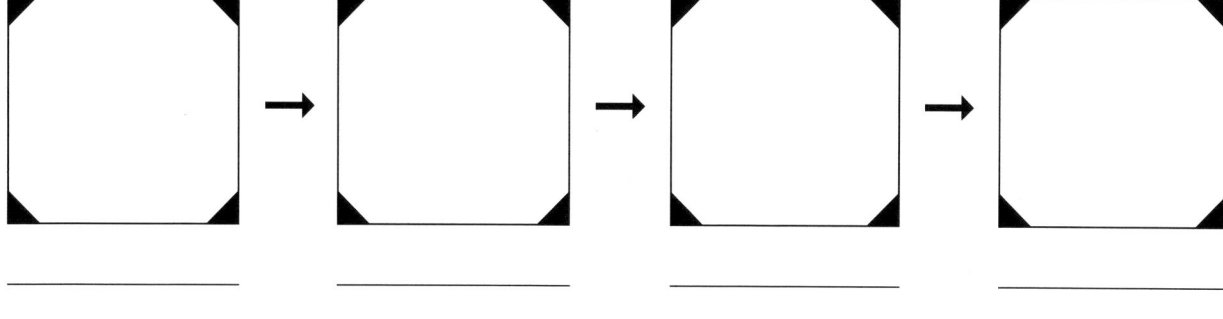

Energy can change from one form to another.

❸ List examples of appliances that change electrical energy into these other forms of energy.

Electrical energy →

Light energy	Heat energy	Sound energy	Movement energy

What is energy?

Fossil fuels

Objective:
- Reads texts and answers questions about fossil fuels.

Teachers notes:
- Initiate a class discussion about fossil fuels. Find out what the pupils already know and record it on the board. Read the information with the class first. Pupils then follow the instructions for Question 2 and complete the note-taking exercise.

- Note taking and writing in their own words can often be a difficult task for many pupils. Writing notes and summarising them is essential for research and project writing, especially in secondary school.

- Help pupils who are having difficulty by asking them to explain to you what they have read. This will help them to organise the information in their heads.

- Pupils read their summaries to the class.

Publishing/Display ideas:
- Pupils find or draw pictures of appliances that use fossil fuels to run. They glue them under one of the three headings: oil, natural gas, coal.

Answers:
Teacher check.

Additional activities:
- Research one of the processes used to find and retrieve a fossil fuel (e.g. oil, natural gas or coal). Present a report to discuss the stages involved.

- What is acid rain? What does it have to do with the burning of brown coal? Find out more about the pollution created by the use of fossil fuels.

Fossil fuels

What is energy?

1. Read the text about fossil fuels. Underline or highlight the keywords, phrases and facts. (Remember: Keywords are the most important words.)

The world's greatest form of energy is solar energy from the sun. However, humans have learnt to release large amounts of energy by burning fossil fuels. The burning of these fuels at power stations gives us the energy for our homes, schools and businesses. We use many kinds of devices to extract the energy from fuels so we can use it to do work, such as heating our homes or cooking our food.

There are three main forms of fossil fuel: coal, oil and natural gas.

These fuels were formed between 100 and 300 million years ago, before and during the age of the dinosaurs. During this time, plants and animals decomposed under tonnes of rock and ancient seas. Eventually, the seas receded and left dry land with layers of fossilised substances underneath it. During the millions of years that passed, the dead plants and animals slowly formed fossil fuels. Different types of fuels were formed, depending on:

- the combination of animals and plants
- how long they were buried
- what the temperature and pressure was when they were decomposing.

Some layers turned into a hard, black, rock-like substance called coal, others a thick, black liquid called oil or petroleum, and some became a natural gas.

Fossil fuels are generally found below ground and each is extracted from the earth in a different way. Coal is mined, while oil and natural gas are pumped to the surface by rigs and transported to oil refineries or storage tanks.

Plastics, fertilisers and even some of the clothes we wear are made from oil. Oil refineries 'split' the oil to form products like petrol, aviation fuel and oil to burn in power plants to produce electricity.

Natural gas can be piped to our homes, schools or businesses to provide heat for warmth or cooking.

Fossil fuels are relatively cheap to produce but they are not renewable. This means they cannot be remade or replaced once they have been used. Also, when they are burned, they produce a lot of pollution as a by-product. Because of this, we must look to alternative power sources, including solar, wind or water (hydro) power, which are all renewable power sources and can be used over and over again!

2. Ask yourself these questions:
- Can I read all the words? ☐
- Do I know the meaning of all the words? ☐
- Do I understand the facts? ☐

3. On a separate sheet of paper, use your highlighted keywords to write notes about fossil fuels. Notes do not need to be in full sentences and can be written in point form (• bullet points).

4. Read through your notes about fossil fuels.
- Do they make sense? ☐
- Do you understand them? ☐

5. Use your notes to write a summary of what you know about fossil fuels. Look away from your notes when you are writing and write in your own words. Good luck!

What is energy?

Renewable energy

Objectives:

- Reads texts and answers questions about renewable energy.
- Conducts research about alternate power sources.

Teachers notes:

- Pupils read the text at the top of the page then use a variety of sources to find the information to answer the questions.

  NOTE: Many useful websites are available to find alternate power sources. Some include:

  www.renewableenergy.com
  www.greenpower.co.uk
  www.rnp.org/renewtech

Answers:

1. *Solar power*
   Solar power can work in two main ways:
   - Solar cells convert light into electricity. Many calculators are powered by solar cells, as are satellites, satellite TV, telephones and the Internet.
   - Solar water heating uses heat from the sun to heat water in glass panels on the roof.

   Wind power
   - The wind blows a propeller around which turns generators to produce electricity. The propellers are usually situated in groups called 'wind farms'. Boats and caravans may have small wind generators.

   Hydroelectric power
   - A dam collects water. Water flows through tunnels in the dam to turn turbines and drive the generators to produce electricity.
2. Geothermal energy, biomass energy, tidal energy, wave energy.

Equipment/Materials required:

- Access to the library and the Internet.

Publishing/Display ideas:

- Pupils choose one of the renewable energy sources listed and create an information poster about it. Posters should be informative, accurate and attention-grabbing. Display the posters.

Additional activities:

- In groups, pupils discuss ways they can educate people about the benefits of renewable energy sources. Pupils discuss how to change people's behaviours. What has been successful in the past? What is needed now?

What is energy?

Renewable energy

We are making ever-increasing demands on the Earth's resources to meet our basic survival needs and to support our modern lifestyle. In the past, we thought we could use all the things around us and they would continue to be there in the future. Now we realise we were wrong and that some of these resources are becoming harder to find or are in danger of running out altogether. We need to conserve our natural resources, not just for today but for our future.

There are two main groups of natural resources:

- Renewable resources

Renewable resources are those that are able to reproduce naturally, such as air, water, forests, soil and animals. Of course, care and conservation are still needed to ensure these resources continue to survive.

- Non-renewable resources

Once non-renewable resources have been removed and used, they cannot be replaced. We burn fossil fuels (coal, oil and natural gas) to use as an energy source for our homes, schools and businesses. Fossil fuels produce a lot of pollution as a by-product, polluting the renewable resources (such as the air).

We must look for other energy sources, including solar power (sun), wind power and hydropower (water), which are renewable resources.

1. Use a dictionary, the Internet, school library and your own knowledge to explain how these alternative power sources work. Include a picture or diagram in the boxes.

SOLAR

_____
_____
_____

WIND

_____
_____
_____

HYDROELECTRIC

_____
_____
_____

2. List at least three other alternative power sources. _____

_____

What is energy?

Energy in the home

Objectives:

- Completes an energy survey of home appliances.
- Lists energy-saving tips for appliances.

Background information:

- As we have become more technologically advanced and environmentally aware, we have begun looking towards creating communities and lifestyles which have a minimal impact on our natural environment.

Teachers notes:

- The main source of energy in our homes is electricity. Other sources include natural gas, solid fuel (wood and coal) and solar energy. Pupils can complete the activity at home if necessary to accurately identify the energy source used by an appliance.

- Pupils can work in small groups to generate ideas for conserving energy with the appliances listed. Direct pupils to consider how each type of energy can be conserved with various appliances; for example, turn off the electric blanket once the bed is warm, or simply use extra blankets.

Equipment/Materials required:

- Pupils may need to take their sheets home to list appliances and determine their energy source.

Answers:

1. Answers will vary.
2. **Cooker:** cover pots or pans to boil or steam faster; turn off the oven ten minutes before food is ready; start cooking with the heat high and lower it when the food starts to boil; preheat the oven only when necessary; bake with glass dishes (they use less heat); use a microwave when possible (they use half the power of a normal oven).
   Central heating: set temperature higher in winter and lower in summer; keep the house closed to the outside and close off unused rooms; wear an extra layer of clothes rather than turning up the heat.
   Lighting: replace bulbs with bulbs of a lower wattage; use fluorescent bulbs where possible; keep light bulbs and fixtures clean.
   Dishwasher: wash only full loads; don't use the dishwasher's dry cycle.
   Refrigerator: clean the coils on the back of the refrigerator (if fitted) at least once a year; check the door seals to see if air is entering; open the door for as short a time as possible.
   Washing machine: wash with warm or cold water and rinse with cold; wash and dry full loads; don't overload the machine.
   Clothes dryer: don't overload the machine; clean the lint trap after every load; dry two or more loads in a row to use the residual heat from the previous load.

Publishing/Display ideas:

- Pupils use their ideas from Question 2 to create a 'Household energy-saving tips' poster. Use large sheets of paper, magazines, Internet access, the library, and glue and drawing materials to create an informative and eye-catching poster.

Additional activities

- Collate the energy-saving hints suggested by the pupils to form a class book or display chart.

- Using the pupils' worksheets, make class tallies of the total number of appliances, according to the energy source or what each is used for. These results can displayed in a bar or pie graph.

Energy in the home

What is energy?

The main source of energy in our homes is electricity. Other sources of energy include natural gas, solid fuel (wood and coal) and solar energy. We use these energy sources to produce heating, cooling, motion (in fans and saws, for example), light, sound, pictures and to provide many other daily needs.

1. List ten appliances in your home which need external energy to operate (i.e. are not hand-operated) and complete the table below.

Appliance	Energy from					Used for						
	Electricity	Gas	Solar	Solid fuel	Other	Heating	Cooking	Cleaning	Cutting	Entertainment	Communicating	Other

2. In the box below, write some 'energy-saving tips' for the appliances shown.

Cooker
_____
_____
_____
_____

Refrigerator
_____
_____
_____

Central heating
_____
_____
_____

Dishwasher
_____
_____
_____

Washing machine
_____
_____
_____

Lighting
_____
_____
_____
_____

Clothes dryer
_____
_____
_____

What is energy?

Changing energy

Objectives:
- Reads texts and answers questions about action and stored energy.
- Makes and tests an 'energy changer'.

Teachers notes:
- Pupils work individually, in pairs or in small groups to create their energy changers, depending on the quantity of materials available.

Equipment/Materials required:
- Rubber band, old-style cotton reel, tape, match, washer, pencil.

Publishing/Display ideas:
- Find photographs or pictures of people and things depicting stored and action energy and display them around the classroom.

Answers:
1. As the pencil is turned, the rubber band is twisted and tightened—stored energy. When the band unwinds, it releases the stored energy as action energy and spins or 'crawls' across the floor.
2. Teacher check.
3. (a) The energy changer is a device that changes stored energy into action energy.
   (b) Holding a rubber band stretched between your fingers is an example of stored energy.
   (c) Letting go so it is flung away is an example of action energy.
4. (a) S   (b) A
   (c) A   (d) A
   (e) S   (f) A
   (g) S   (h) A

Additional activities:
- Energy is required to wind up the rubber band of the energy changer. Where does this energy come from?
- Why is sound considered to be an 'action' energy? Discuss.

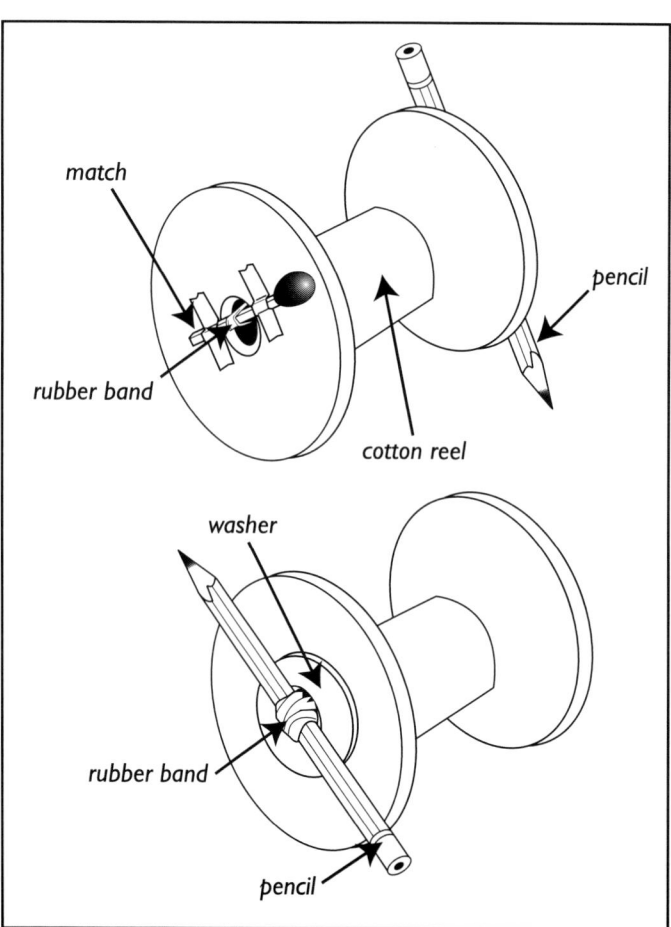

Investigating renewable energy and sustainability — www.prim-ed.com — Prim-Ed Publishing

What is energy?

Changing energy

Energy is something that we need to make things happen. Energy can exist in many different forms. Whatever form it takes, energy can be used to make things move.

There are two main forms of energy:

- **action** energy
  Anything that is moving. Action energy is also known as **kinetic** energy.

- **stored** energy
  Anything that is stretched, compressed (squashed) or held above the ground. Stored energy is also known as **potential** energy.

Energy can change from one form to another but can never be lost or destroyed.

Make an energy changer

You will need:
- rubber band
- old-style cotton reel
- tape
- match
- small washer
- pencil

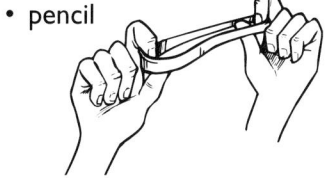

What to do:
1. Thread the rubber band through the cotton reel.
2. Carefully snap the match so it stays in one piece, shorter than the cotton reel's diameter. Place it through the loop of the rubber band at one end of the cotton reel.
3. Tape the match to the cotton reel so it is unable to move.
4. Thread the other loop of the rubber band through the washer.
5. Place the pencil through the loop on the outside of the washer.
6. Use the pencil to carefully wind up the rubber band.
7. Place the energy changer on the floor.

1. What happened?

_____
_____
_____
_____
_____

2. Diagram with labels

3. Fill in the missing words.

(a) The energy changer is a device that changes _____ energy into _____ energy.

(b) Holding a rubber band stretched between your fingers is an example of _____ energy.

(c) Letting it go so it is flung away is an example of _____ energy.

4. Write **A** for action energy or **S** for stored energy in the boxes next to each example.

(a) Rock balanced on the edge of a cliff. ◯   (b) Kicking a football. ◯

(c) Riding a bicycle. ◯   (d) Apple falling from a tree. ◯

(e) Waiting at the top of a roller-coaster ride. ◯   (f) Blades of a windmill turning. ◯

(g) River held behind a dam wall. ◯   (h) Pistons moving up and down in an engine. ◯

What is energy?

Types of energy

Objectives:
- Illustrates the different types of action and stored energies.
- Lists types of energy transfers.

Teachers notes:
- Explain to the class that another name for stored energy is potential energy. Ask pupils for a definition of the word 'potential'. To help the class understand potential energy, use the example of a person pulling back a bow when using a bow and arrow. This is quite a difficult task and the person can probably feel the stored energy in the bow.

Publishing/Display ideas:
- Pupils can transfer their picture ideas from Question 1 onto art paper and create a poster of the different types of energy.

Answers:
1. Teacher check.
2. (a) chemical stored energy → kinetic energy
   (b) electrical energy → light energy + heat energy

Additional activities:
- As a class, discuss how each of the following things has energy: the sun, a falling stone, piece of wood, a battery.
- Think of a major sporting event such as a football or cricket match. In small groups, pupils list the different types of energy found there.

What is energy?

Types of energy

Energy surrounds us in different forms. The two most common forms of energy we use are heat and electricity. Many machines that make our work easier use either heat or electricity or both, such as a clothes dryer.

We also use many other forms of energy in our daily lives.

1. Read about the different types of energy and draw a picture to represent each.

ACTION ENERGY – energy of movement that can be seen (like a speeding car) or not seen (like particles in solids, liquids and gases).	STORED ENERGY – energy that can be stored, ready for action!
Kinetic energy (energy an object has due to movement)	**Gravitational stored energy** (energy stored in an object lifted above the ground)
Light energy (energy from the sun or light bulbs)	**Chemical stored energy** (found in fuels and food)
Heat energy (energy from the sun, friction, burning fossil fuels or electricity through an element)	**Elastic stored energy** (objects that are stretched or compressed)
Sound energy (energy from vibrating air or objects)	
Electrical energy (energy of moving electrons through a conductor such as a wire)	**Nuclear stored energy** (energy in atoms of certain elements like uranium)

2. Energy can be changed from one form to another. Fill in the types of energy produced for each.

(a) _____ stored energy → _____ energy

(b) _____ energy → _____ energy + _____ energy

Solar energy

The sun

Objective:
- Reads texts and answers questions about the sun.

Background information:
- Humans have been relying on the sun's heat and light for millions of years. If the sun was to disappear, animals and plants would not be able to survive on Earth. The rays of the sun bring large amounts of light to Earth. This light is converted to heat energy within the Earth's atmosphere, which holds much of the heat 'in'.

Teachers notes:
- Food chains are not isolated behaviours but part of more complex food webs. However, food chains are an excellent way of developing an understanding of how energy is transferred and of the principle that all organisms need energy to live.
- Pupils may wish to conduct research to answer Question 3, so access to the library and the Internet may be required for this lesson.

Equipment/Materials required:
- Access to the Internet and library.

Publishing/Display ideas:
- Display sunrise and sunset data with corresponding graphs.
- In groups, design and make collages of the sun, and plants and animals in food chains, to display and label.

Answers:
1. – 3.   Teacher check.

Additional activities:
- Create an informative poster explaining why we have seasons.
- Search the Internet for sunrise and sunset times where you live. Look at a globe and choose a country that is on the opposite side of the world to you. Based on the data collected, how does the length of their day compare to yours?

Solar energy

The sun

The sun is a star around which the Earth and all the other planets in our solar system revolve. For billions of years, the sun has given out huge volumes of energy that supports all life on Earth. Green plants use the sun's energy to make their own food, which they store in their leaves. They do this using a process called photosynthesis.

Food chains show how energy is transferred between plants and animals in the environment. The sun is the source of the energy and begins every food chain.

1. Explain what is happening in the food chain below.

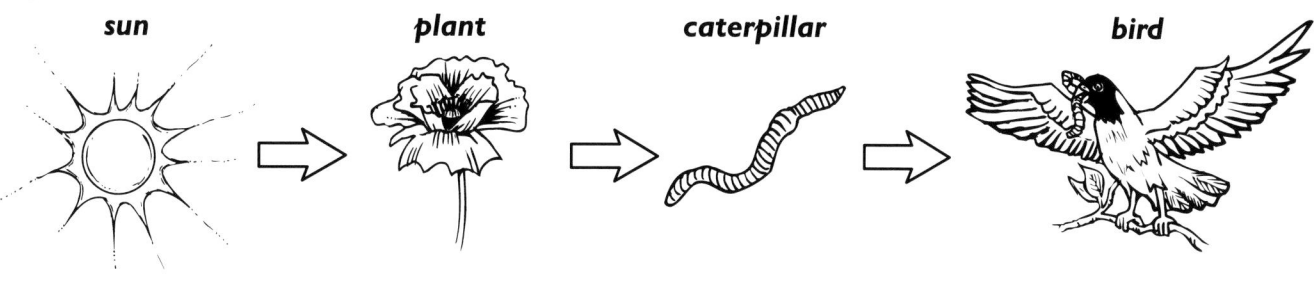

_____
_____
_____

2. Draw and label a food chain that can be found in your local environment.

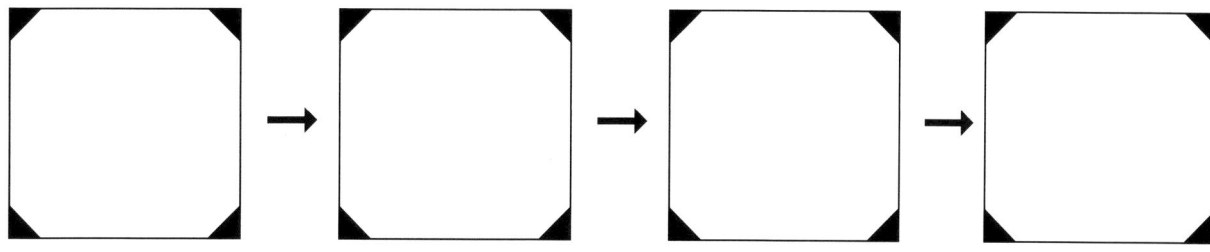

Energy from the sun is called solar energy and most of this reaches the Earth as light. To be useful to us, we need to be able to convert this light energy into heat and electricity—the two forms of energy we use the most.

3. How do you think a greenhouse or a solar hot water heater works? Choose one and write your ideas below. Include a diagram with your explanation.

Greenhouse ☐   Solar hot water heater ☐

_____
_____
_____
_____
_____
_____
_____

Solar energy

Objective:
- Reads texts and answers questions about solar energy.

Teachers notes:
- Pupils read the text at the top of the page then discuss the questions with a partner before completing the answers.

Equipment/Materials required:
- Bring items that are powered by solar cells to class for display.

Publishing/Display ideas:
- Display photographs of satellites orbiting the Earth.
- Create a table of items that are powered by solar cells.

Answers:
1. Solar hot water systems, making electricity.
2. (a) A large, flat, metal box is installed on the roof of a home. It contains pipes that are connected to the home's plumbing. The metal lining and the pipes are heated when the sun shines on the box. This heat is carried away by the water flowing through the pipes. The now hot water is stored in an insulated tank for later use.
   (b) Lower power bills; less reliance on fossil fuels so less impact on the environment; solar energy is a renewable energy source.
3. (a) Solar cells are cells made from thinly-sliced silicon that can convert solar energy directly to electrical energy.
   (b) Water pumps, calculators, watches and signs, satellite TV, weather forecasting, the Internet and much more.
4. Solar energy is a very practical choice for countries with warmer climates and lots of sunshine, such as Australia.

Additional activities:
- Imagine life without solar cells. Write a creative story where all of the satellites in orbit disappeared so satellite TV and the Internet no longer existed.
- Research to find more about the properties of silicon. Present an information poster about silicon.

Solar energy

There is a great amount of power in the sun's rays that reach the Earth. This power is called solar energy and is measured in watts per square metre.

Solar hot water heating

Solar energy can be used to heat water in your home by way of a solar hot water heating system. A large, flat metal box is installed on the roof of a home. It contains pipes that are connected to the home's plumbing. The metal lining and the pipes are heated when the sun shines on the box. This heat is carried away by the water flowing through the pipes. This now hot water can be stored in an insulated tank for us to use when we have a shower or do the washing. These systems can supply some or even all of a home's hot water, making power bills considerably lower!

Making electricity

Sunlight can be converted into electricity by using a photovoltaic cell, also called a solar cell. These cells are made from thinly sliced silicon. A single cell can only produce tiny amounts of electricity, so many need to be joined together. Water pumps, calculators, watches and signs can be powered using solar cells. On a sunny day, one square metre of these silicon cells, also known as a solar panel, can run a 100 watt light bulb.

Solar cells were originally developed to provide electricity for satellites that orbit the Earth. These satellites allow us to have satellite TV, weather forecasting, the Internet and much more.

Solar energy is a very practical choice in countries with warmer climates and lots of sunshine, such as Australia.

To make electricity at night or on cloudy days, however, the energy must be stored for later use.

Answer these questions.

1. List the two main uses of solar energy.

_____

2. (a) In your own words, explain how a solar hot water system works.

_____
_____
_____

(b) Name two benefits of using solar energy to heat water in your home.

• _____
_____

• _____
_____

3. (a) In your own words, explain how a solar cell works.

_____
_____

(b) List at least five uses of solar cells and draw one of them.

_____
_____
_____

4. Explain why solar power is a sensible choice for an energy source in a country such as Australia.

_____

Solar energy

Uses of solar energy

Objective:
- Plans, designs and creates a poster about the uses of solar energy.

Background information:
- Uses of solar energy include:
  - greenhouse
  - solar hot water heaters
  - solar ovens and cookers
  - home heating
  - solar torches
  - solar calculators
  - watches
  - portable construction signs
  - emergency telephones
  - solar showers
  - solar-powered toys (racing cars and speedboats)
  - battery-operated toys (battery re-energised by solar-panelled charger)
  - solar electricity to power homes, sailboats, recreational vehicles, businesses

Teachers notes:
- Pupils could work individually, with a partner or in a small group for this activity.
- Organise research time in the library or computer room. Alternatively, have available to the pupils a number of books about solar energy and computers with specific websites bookmarked.
- Points to consider for a successful poster:
  - Brief, concise information written in own words (bullet points are effective).
  - Overall attractiveness.
  - Attention-grabbing titles.
  - Clear, simple bold design that gives viewers a quick overview of the topic without a lot of reading.
  - Colour (bright to draw attention).
  - Consideration given to art style/design; for example, art style (media), type of graphics, placement of pictures.

Equipment/Materials required:
- Materials for creating the posters such as drawing equipment, large sheets of plain and coloured card, scissors and glue.

Publishing/Display ideas:
- Display completed posters for other pupils in the school to read. For example, display them on the school noticeboard, in the library or on a wall along a thoroughfare.

Answers:
Teacher check

Additional activities:
- Find out how the school and the community use solar power for energy.
- What are the disadvantages and advantages of using solar power as an energy source? Discuss.

Solar energy

Uses of solar energy

Every day, the sun pours out an enormous amount of energy. How do we use this energy on Earth?

1. Design and create an information poster about the uses of solar energy. Research the topic and record notes for your poster in the sections below. Keep a record of the books and websites you use.

Design brief/Purpose _____

Solar energy uses – key research points

Diagrams/Art

Poster layout/design

Catchy titles

Information sources

2. Use your plan to complete your poster on a large sheet of card or heavy paper. Make sure your poster is informative, has accurate facts and vocabulary and is eye-catching!

Solar energy

Build your own solar oven

Objectives:

- Follows a procedure to construct a solar oven.
- Explains how and why a solar oven works.

Teachers notes:

- Pupils work in groups, checking that they have all the materials from the 'What you need' list. Groups follow the instructions to make a solar oven.

Equipment/Materials required:

- 2 polystyrene cups, large family-size yoghurt pot, tissue paper, baking foil, black paper, large sheet of stiff paper, plastic food wrap, adhesive tape, sliced apple, thermometer, watch/timer

Publishing/Display ideas:

- Display the pupils' solar ovens. Use a digital camera to photograph the pupils making their ovens and display photos on a board near the ovens.

Answers:

1. – 2. Teacher check.
3. The solar oven traps the sun's heat within it. The foil inside the oven reflects the sun like a mirror. The black cone shaped paper directs the light (and heat) onto the apple, cooking it. The inner cup is lined with black paper which helps to absorb heat; the food wrap prevents the heat from escaping.

Additional activities:

- 'If solar energy is free, why don't we just power everything using solar energy?' Discuss.
- Pupils choose another piece of food they would like to cook and design a new type of solar oven that will fit this food. (For example, a shoe box or cereal carton could be used instead of the yoghurt pot.) Pupils design the new oven, construct it and make predictions regarding the cooking time of the food. Pupils test their oven and predictions.

Solar energy

Build your own solar oven

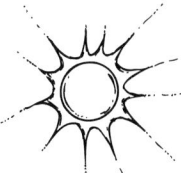 The sun's rays bring large amounts of light to Earth. This light is converted to heat energy within the Earth's atmosphere, which holds much of the heat 'in'. Energy from the sun is called solar energy. This energy can be trapped to warm our surroundings, our homes and ourselves.

What you need:

- 2 polystyrene cups
- large family-size yoghurt pot
- tissue paper
- baking foil
- black paper
- large sheet of stiff paper
- plastic food wrap
- adhesive tape
- sliced apple
- thermometer
- watch/timer

What to do:

1. Line one polystyrene cup with black paper and place the sliced apple in it. Tightly cover the top with plastic wrap.
2. Cover one side of the sheet of paper with foil. Wrap it around the cup and tape it in place, with the foil on the inside.
3. Place the cup and black paper cone into the second cup. Place this inside the yoghurt pot. Use the tissue paper to fill in the gaps between the cup and pot.
4. Place your solar oven in the sunshine and angle it directly towards the sun.

1. In your group, decide how you will measure the changing temperature in the solar oven. Write your ideas below. *(Note: You may wish to alter the solar oven design slightly.)*

_____

_____

_____

2. Construct a table of results to record the time, temperature and changes in the apple. Pictures can also be included in the table to show how the apple changes.

3. Explain how and why the apple is cooking in the solar oven.

_____

_____

_____

_____

_____

Prim-Ed Publishing   www.prim-ed.com   Investigating renewable energy and sustainability

Solar energy

Solar energy – advantages and disadvantages

Objective:
- Lists the advantages and disadvantages of using solar energy as a power source.

Background information:

ADVANTAGES
- Solar energy is free.
- Requires no fuel.
- Produces no waste or pollution.
- Can be used in places with warm climates for electricity production (especially places where obtaining electricity is difficult).
- Good for low-power uses such as battery chargers and garden lights.

DISADVANTAGES
- Unreliable in areas with cooler climate (such as the UK and Ireland).
- Expensive to build solar power stations.
- Solar cells are expensive in relation to the electricity they are able to produce in their lifetime. (However, technology is rapidly improving this.)

Equipment/Materials required:
- Access to the Internet and library

Answers:
1. Teacher check.
2. See background information.
3. Teacher check.

Teachers notes:
- Ensure the pupils have adequate time for discussion to list their own suggestions for the advantages and disadvantages of solar power before they use the Internet and library for further research.
- To save time and fuss, the websites listed could be bookmarked.
- Although pupils work in a group to complete Questions 1 and 2, Question 3 should be completed individually. Some pupils may need assistance to transfer their lists into a paragraph that summarises both sets of ideas.
- Remind pupils to conclude the paragraph with a personal comment about the potential of solar power to become a major energy source.

Publishing/Display ideas:
- Use coloured A3 (or larger) sheets of cardboard to collate each group's ideas. Display them.

Additional activities:
- Pupils read their summarising paragraph to the class or their group. Keep a record of how many pupils felt solar power could become a mainstream energy source and how many did not. Present and discuss the data.

Solar energy

Solar energy – advantages and disadvantages

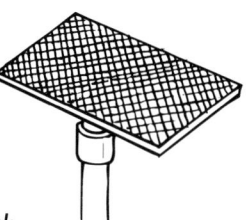

1. In a small group, discuss and list the advantages and disadvantages of using solar power as an energy source.

2. When the discussion concludes, use the library and the Internet to research the topic further. Enter keywords into a search engine such as *Google*. Useful websites include:

   www.renewableenergy.com   www.greenpower.co.uk   www.rnp.org/renewtech

Solar power	
Advantages	**Disadvantages**

3. Write a paragraph that summarises your findings. Conclude with a personal comment stating your opinion about solar power and its potential to become a major energy source. Continue on the back of this sheet if necessary.

_____
_____
_____
_____
_____
_____

Wind energy

Objective:
- Reads text and answers questions about wind energy.

Teachers notes:
- Ask the class if they have ever been 'bowled over' or 'propped up' by strong winds. Discuss extreme weather conditions such as tornadoes and cyclones. Introduce the concept that wind has energy.

Publishing/Display ideas:
- Pupils research the inside of a wind turbine, draw their own labelled diagram and add notes explaining each part. Display diagrams.

Answers:
1. Pump water for crops, grind grains to make flour, pump water to make farmland, create electricity.
2. The sun warms the air unevenly, so some patches are warmer than others. When the warm air rises, cooler air from surrounding areas rushes in to fill the space, creating wind.
3. Wind is captured by wind turbines —tall towers with large propellers at the top. Small, light generators behind the propellers change the wind energy to electricity.
4. By the coast, on hill tops and in gaps between mountains.
5. renewable energy source; produces no air pollution; has relatively little impact on the environment

Additional activities:
- Pupils research the use of dykes in the Netherlands and present a brief oral report.
- Challenge the class to find out the minimum wind speed required for a wind turbine to generate electricity.

Wind energy

Have you ever watched a documentary or news bulletin showing the powerful force of a cyclone or tornado? Such winds possess an enormous amount of energy and can have devastating effects on property, people and the environment.

Humans have been harnessing wind energy for thousands of years. It is believed the Chinese have been using wind to pump water for crops for 4000 years. Windmills have been used in Europe for centuries to turn heavy granite stones to grind grains such as wheat, barley and oats to make flour. Traditionally, the Dutch used windmills to pump sea water from low-lying estuaries, creating land that could then be farmed.

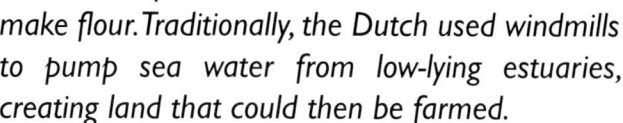

What makes the wind?

The sun warms the air unevenly, so some patches are warmer than others. When the warm air rises, cooler air from surrounding areas rushes in to fill the space, creating wind.

Wind can be 'captured' by wind turbines—tall towers with large propellers on the top—and the energy harnessed. This moving energy (kinetic energy) can be changed into electricity (electrical energy) using a generator. The generators are small and light and can be housed at the top of the tower.

The stronger the breeze, the more electricity that can be produced, so the wind turbines need to be placed in windy areas such as by the coast, on hilltops and in gaps between mountains. A number of towers are often placed together, creating a 'wind farm'. The more towers and the more wind, the more electricity that can be made.

Wind energy is a renewable energy source that produces no air pollution and has relatively little impact on the environment.

Answer these questions.

1. List three uses of wind energy.

- _____
- _____
- _____

2. In your own words, explain why wind occurs. Include a diagram.

_____

_____

_____

3. Fill in the gaps to explain how wind energy is changed to electrical energy.

Wind is captured by wind _____ — tall towers with large _____ at the top. Small, light _____ housed behind the propellers change the wind energy to _____.

4. Where should wind farms be situated for the best results?

_____

_____

5. On the back of this sheet, list three benefits of using wind power as an energy source.

Wind energy

Wind farms

Objective:
- Uses a variety of resources to plan and write a report about wind farms.

Background information:

A wind farm is a collection of wind turbines in the same location. Wind turbines are used to generate electricity. Wind is 'captured' by wind turbines and the energy harnessed. This moving energy (kinetic energy) can be changed into electricity (electrical energy) using a generator. The generators are small and light and can be housed at the top of the tower.

The stronger the breeze, the more electricity that can be produced, so the wind turbines need to be placed in windy areas such as by the coast, or on hilltops and in gaps between mountains. Wind farms can be positioned on land or offshore. In Europe, offshore farms are common. They are just starting to be implemented in the United States.

Germany has the greatest number of wind farms in the world and the largest offshore wind farm.

Wind energy is a renewable energy source that produces no air pollution and has relatively little impact on the environment.

Teachers notes:
- Review report writing with the class. Review how to write in the third person.

Equipment/Materials required:
- Access to the library and Internet, word processor and art paper.

Publishing/Display ideas:
- Published reports, accompanied with appropriate illustrations and diagrams, could be displayed for other pupils to read and compare.

Answers:
Teacher check.

Additional activities:
- Finished reports could be read orally by pupils in small groups or to the whole class.

Wind energy

Wind farms

1. Use the Internet, library and your own knowledge to plan an information report about wind farms. Include accurate facts and vocabulary. Research to find where wind farms are located in your area/county/country.

Title

Classification (A general statement about the topic)

Description (Divide the description into sections; e.g. What are wind farms? How do they generate electricity? Where are they found and why? Why they are used?)

- Reports are written in the third person; e.g. The turbines; They work by ...

- Reports use factual language rather than imaginative; e.g. The small, light generators are housed ...

- Reports use technical vocabulary and subject-specific terms; e.g. propeller, volume of air ...

Conclusion (A final comment about the subject of the report; it may contain a personal opinion.)

- Reports have information organised into paragraphs.

2. Write your report on a separate sheet of paper. Edit your work and present a finished copy with diagrams and pictures.

Wind energy

— Make an anemometer —

Objectives:
- Builds an anemometer.
- Writes a set of instructions clearly explaining how to build an anemometer.

Background information:
- An anemometer is used by meteorologists to measure wind speed. The 'cups' catch the wind and spin. The speed of the rotation is read by a special device that converts it into wind speed.

Equipment/Materials required:
- Yoghurt cups, corks, pen lid, thin and thick dowels, wooden block, adhesive materials.

Answers:
1. – 5. Teacher check.

Teachers notes:
- In groups, pupils construct their model anemometer. Explain that meteorologists use anemometers to read wind speed.
- For Question 2, ask the class to write the steps in order and begin each sentence with a verb. A set of instructions is written as a procedure, with each step beginning with a command verb. For example, instead of 'We carefully pushed the thin dowel into the cork,' write it as 'Push the thin dowel into the cork carefully'.

Publishing/Display ideas:
- Display extreme weather photos or posters. Display a world map. Add research and pictures of areas in the world that experience extreme weather conditions such as cyclones.

Additional activities:
- *What is the difference between a hurricane, typhoon and cyclone?*

    Research to find the answer and record your findings on the back of this sheet.

    Answer: Powerful rotating storms are called hurricanes when they start over the Atlantic Ocean, typhoons when they form over the Western Pacific, and cyclones when they start over the Indian Ocean or Southern Pacific Ocean.

Investigating renewable energy and sustainability

Wind energy

Make an anemometer

An anemometer is used by meteorologists to measure wind speed. The 'cups' catch the wind and spin. The speed of the rotation is read by a special device that converts it into wind speed.

1. Use the materials listed and the diagram to build an anemometer. Add any extra materials you used to the list.

You will need:
- yoghurt cups
- corks
- pen lid
- dowel rods – thick and thin
- wooden block

• _____

• _____

• _____

2. Write a set of instructions that clearly explain how to build an anemometer.

_____
_____
_____
_____
_____
_____
_____
_____
_____
_____
_____
_____

3. Test your anemometer. What happened? How could you improve your anemometer?

_____
_____
_____
_____

4. Watch your anemometer for 30 seconds. Count the revolutions. (Attach a marker to one cup to help you.) If 50 revolutions in 30 seconds = 18 kilometres per hour, estimate how fast the wind is blowing.

☐ kilometres per hour

How did you work it out?

_____
_____
_____

5. Use your anemometer every day for one week. Do you think a wind farm would be a sensible choice for producing electricity in your area? Explain why or why not. Continue on the back of this sheet.

_____
_____

Wind energy

Wind energy – advantages and disadvantages

Objective:
- Lists the advantages and disadvantages of using wind energy as a power source.

Background information:

ADVANTAGES
- Wind is free.
- Wind farms do not require fuel.
- No air pollution.
- Good for rural/remote areas.
- Little impact on land.
- Land beneath can still be used for farming.
- Wind farms can become tourist attractions.

DISADVANTAGES
- Wind can be unpredictable.
- Turbines can be noisy.
- Turbines can kill birds.
- Wind farms may be seen to be unsightly.
- Can affect television reception.

Teachers notes:
- Ensure the pupils have adequate time for discussion to list their own suggestions for the advantages and disadvantages of wind power before they use the Internet and library for further research.
- To save time and fuss, the websites listed could be bookmarked.
- Although pupils work in a group to complete Questions 1 and 2, Question 3 should be completed individually. Some pupils may need assistance transferring their lists into a paragraph that summarises both sets of ideas.
- Remind pupils to conclude the paragraph with a personal comment about the potential of wind power to become a major energy source.

Equipment/Materials required:
- Access to the Internet and library.

Publishing/Display ideas:
- Use coloured A3 (or larger) sheets of cardboard to collate each group's ideas. Display them.

Answers:
1. – 2. See background information.
3. Teacher check.

Additional activities:
- Pupils read their summarising paragraph to the class or their group. Keep a record of how many pupils felt wind power could become a mainstream energy source and how many did not. Present and discuss the data.

Wind energy

Wind energy – advantages and disadvantages

1. In a small group, discuss and list some advantages and disadvantages of using wind power as an energy source.

2. When the discussion concludes, use the library and the Internet to research the topic further. Enter key words into a search engine such as *Google*. Useful websites include:

   www.renewableenergy.com   www.greenpower.co.uk
   www.rnp.org/renewtech

Wind power	
Advantages	**Disadvantages**

3. Write a paragraph that summarises your findings. Conclude with a personal comment stating your opinion about wind power and its potential to become a major energy source. Continue on the back of this sheet.

_____
_____
_____
_____
_____
_____

Hydropower

Moving water

Objective:
- Reads text and answers questions about hydropower.

Background information:
- About 24% of the world's electricity comes from the energy of falling water. Electricity that has been generated by falling water is called hydro-electricity.

Teachers notes:
- Pupils read the text at the top of the page then discuss the questions with a partner before completing the answers.

Publishing/Display ideas:
- Display photographs and pictures of dams and moving water such as Niagara Falls (on the US-Canada border).

Answers:
1. To move stone, to grind grain, run pumps and drive industrial machinery.
2. (a) pressure: application of force
   (b) renewable: naturally occurring and can be replenished (in theory is inexhaustible)
   (c) turbine: a motor in which a vaned wheel is made to turn by wind, liquid, steam, burning fuel etc.
3. Teacher check.
4. It is renewable; no pollution or waste produced.

Additional activities:
- Find out how a fish ladder is used to help salmon and other migratory fish swim upstream when dams have been constructed.
- Discuss possible disadvantages of using hydropower as an energy source.

Hydropower

Moving water

Water is constantly on the move. It falls from the sky as rain, runs down mountains and into rivers and creeks and is washed into the ocean. Over time, the force of moving water down a river can slice through a mountain range and carry millions of tonnes of soil into the ocean. Moving water is one of nature's greatest forces!

Waterwheels

Water has been used as an energy source for thousands of years. The ancient Egyptians used water as a power source to move stone. Waterwheels have been used to grind grain and run pumps for thousands of years and, later, they were used to drive industrial machinery.

Steam power

In the 1700s, a workable steam engine was developed that turned water into steam and used the pressure developed to drive a piston (a cylinder on a rod) up a pipe. The steam, at first, was then lost, until later changes to the engine allowed the steam to be condensed (changed back to a liquid) and the water used again.

Energy that comes from the force of moving water is called **hydropower**. Water from a river is held back behind a dam wall. Here it has stored energy, which is energy waiting to be used. The water falls through huge pipes, gaining energy (movement energy) as it travels to a building with large turbines. The water turns the turbines at high speed. These turbines are connected to generators, which produce electricity.

When water is used as a source of energy it is renewable (unlike fossil fuels such as coal and oil) and no pollution or wastes are produced.

Answer these questions.

1. List four uses of waterwheels.

_____

_____

2. Use a dictionary to find definitions for these terms:

• pressure _____

• renewable _____

• turbine _____

3. Draw diagrams and add notes to explain how moving water can generate electricity.

[ ] → [ ] → [ ]

4. On the back of this sheet, list two benefits of using water power as an energy source.

Hydropower

Hydroelectricity

Objective:
- Completes a cloze exercise about hydroelectricity.

Teachers notes:
- Show the pupils a picture or photograph of people whitewater rafting. Have any pupils ever seen or been whitewater rafting? Discuss. Explain that the river's power forces the boat or raft down the river. Where does the power come from? Describe a wide river flowing to a narrow point. Not all of the river can flow through and so water is stored, waiting. This is called stored energy. It is this energy that pushes the water through the narrow opening with a great force.

- Similarly, water in a reservoir that has come from a river is waiting behind a dam wall, wanting to flow through. The energy stored here can be harnessed and used to make electricity.

- Display pictures and photographs of hydropower plants and people whitewater rafting. Pupils add notes to the photographs explaining what is being displayed.

Answers:
1. plant
2. dam
3. energy
4. penstock
5. pressure
6. turned
7. generator
8. voltage
9. lines
10. homes
11. outflow

Additional activities:
- Use the Internet to find the names and locations of hydropower plants. Mark the locations on a map of the world. Where are the most hydropower plants found? Why do you think this is?

Hydroelectricity — Hydropower

Use the diagram and the words listed below to complete the passage about how a hydropower plant works.

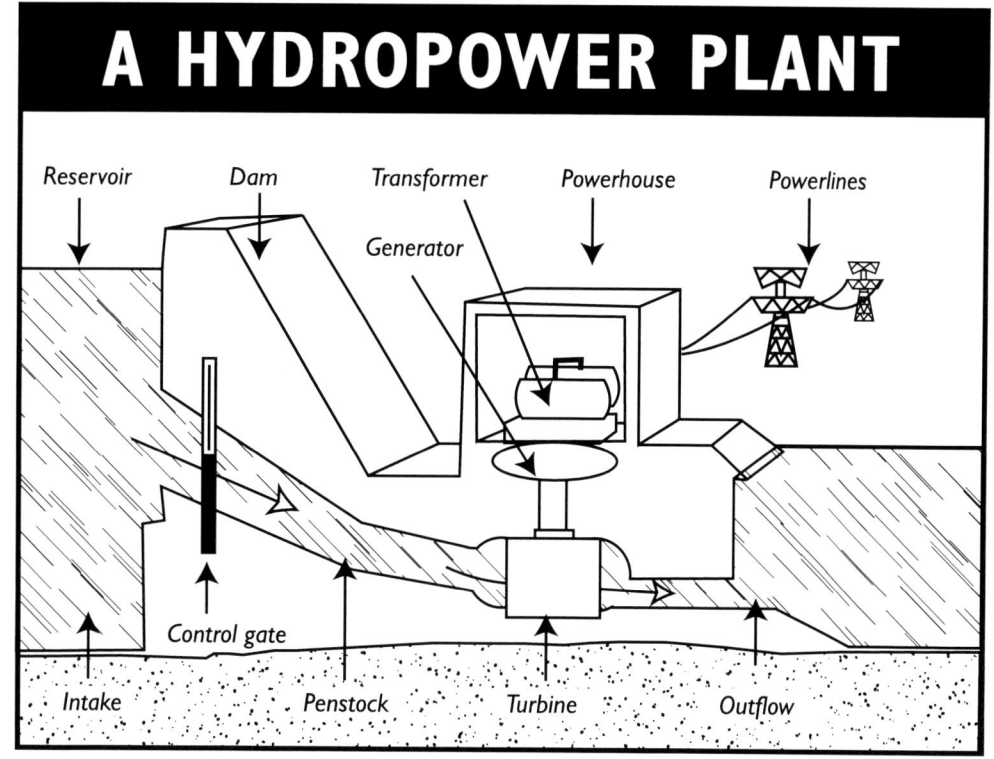

dam	generator	outflow	turned
voltage	lines	pressure	plant
penstock	energy	homes	

A hydropower _____¹ converts the water's energy into electricity.

Water from a river is collected in a reservoir by a _____² that holds the water back.

This water is storing _____³ because it wants to flow through the dam. When the water is needed, a gate is opened and gravity pulls the water through a pipeline called the _____⁴.

The water builds up _____⁵ in the pipe and flows to the turbine. Large blades in the turbine are _____⁶ by the water pressure, which then turns magnets in the _____⁷.

These giant magnets rotate past copper coils, which produce a current. The transformer converts the current to a higher-_____⁸ current.

Out of every powerhouse are power _____⁹ which direct the electricity to our _____.¹⁰ The used water, called _____¹¹, is carried through pipelines and re-enters the river.

Hydropower

Power of water investigation

Objective:

- Follows a procedure to investigate the power of water.

Background information:

- Newton's Third Law states that for every action there is an equal and opposite reaction. The forces from the water pouring out of the holes push the carton in the opposite direction. The carton spins. The more holes, the more force and the faster the carton turns.

- This experiment shows that water which falls from a height and is forced through pipes has energy. In a hydropower plant, water is forced through pipes. This energy spins turbines, which can then be connected to a generator to make electricity.

Teachers notes:

- Remind pupils of the tips for drawing scientific diagrams:
    - Use a sharp pencil.
    - Use a ruler for straight lines.
    - Use all the space you have.
    - Label all parts.
    - Rule a line from each piece of equipment to your label.
    - Give your diagram a title.

Equipment/Materials required:

- 2-litre milk or juice carton, string, nail, water, jug, masking tape.

Publishing/Display ideas:

- Take digital photographs of the pupils conducting the experiment and display them.

Answers:

Teacher check (see background information).

Additional activities:

- Pupils imagine they are a drop of water and write an imaginative recount of the droplet's journey, beginning as part of a cloud and eventually becoming part of a hydroelectric power plant.

Hydropower

Power of water investigation

What you need:
- 2-litre milk or juice carton
- nail
- jug
- string
- water
- masking tape

What to do:

1. Use the nail to carefully punch a hole in the centre of the top part of the milk carton.
2. Push string through the top hole and tie it. Make sure the milk carton will hang from the string.
3. Carefully punch holes in the bottom right corner of each *side* of the milk carton (four in total).
4. Using masking tape, tape each hole you made.
5. Open the spout of the milk carton.
6. Go outside and find somewhere sensible to hang the milk carton (such as a low tree branch).
7. Pour water in the jug and use it to carefully fill the milk carton with water.
8. Pull off the tape from one corner of the milk carton. What happens?

_____

_____

9. Pull off the tape from two corners on the carton that are opposite each other. What happens?

_____

_____

10. Finally, pull off the last piece of masking tape. What happens? Draw and label diagrams to aid your explanation.

_____

How do you think this experiment is similar to water in a hydropower plant? Discuss your ideas with your group and record your findings below and on the reverse if necessary.

_____

_____

_____

Hydropower

Hydropower – advantages and disadvantages

Objective:

- Lists the advantages and disadvantages of using hydropower as a power source.

Background information:

ADVANTAGES

- Once the dam has been constructed, the energy is acquired cheaply.
- No pollution or waste produced.
- Reliable energy source (compared to other renewable energy sources such as solar and wind).
- Water can be stored and used during drought.
- Some dams have more than one purpose—used for irrigation and flood control.
- Electricity generated constantly.

DISADVANTAGES

- Dams are expensive to build.
- Building a dam floods a large area upstream, which causes problems for animals and plants (especially for fish and other creatures which normally migrate upstream).
- The quality of the water downstream of the dam is affected, which has an impact on animals and plant life.
- When the reservoir is filled, areas of farmland and forest can be covered by water.
- Finding a site for a dam can be a problem and it most likely will have an impact on people and the environment.

Teachers notes:

- Ensure the pupils have adequate time for discussion to list their own suggestions for the advantages and disadvantages of hydropower before they use the Internet and library for further research.
- To save time and fuss, the websites listed could be bookmarked.
- Although pupils work in a group to complete Questions 1 and 2, Question 3 should be completed individually. Some pupils may need assistance to transfer their lists into a paragraph that summarises both sets of ideas.
- Remind pupils to conclude the paragraph with a personal comment about the potential of hydropower to become a major energy source.

Equipment/Materials required:

- Access to the Internet and library.

Publishing/Display ideas:

- Use coloured A3 (or larger) sheets of cardboard to collate each group's ideas. Display them.

Answers:

1. – 2. See background information.
3. Teacher check.

Additional activities:

- Pupils read their summarising paragraph to the class or their group. Keep a record of how many pupils felt hydropower could become a mainstream energy source and how many did not. Present and discuss the data.

Hydropower

Hydropower – advantages and disadvantages

1. In a small group, discuss and list some advantages and disadvantages of using hydropower as an energy source.

2. When the discussion concludes, use the library and the Internet to research the topic further. Enter keywords into a search engine such as *Google*. Useful websites include:

www.renewableenergy.com   www.greenpower.co.uk
www.rnp.org/renewtech

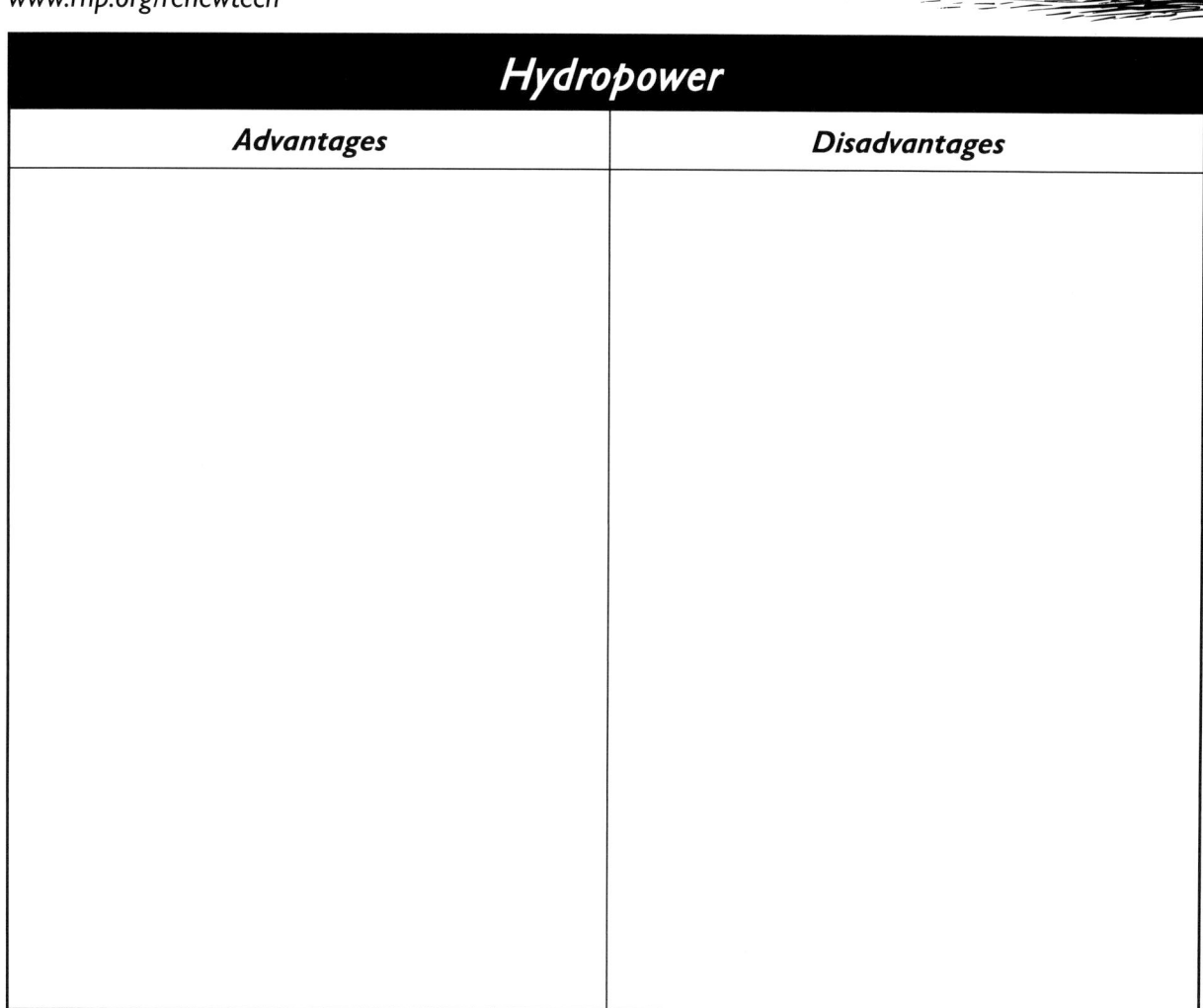

Hydropower	
Advantages	**Disadvantages**

3. Write a paragraph that summarises your findings. Conclude with a personal comment stating your opinion about hydropower and its potential to become a major energy source. Continue on the back of this sheet if needed.

_____
_____
_____
_____
_____

Alternative energy sources

Wave energy

Objective:
- Reads texts and answers questions about wave energy.

Background information:
- As wind travels across the surface of the ocean, it pushes against the water and the energy in the wind is absorbed by the water, causing waves.
- When wave energy is harnessed it can be used:
  - to generate electricity
  - for desalination
  - for pumping water into reservoirs.
- In Scotland, the 'LIMPET' (Land Installed Marine Power Energy Transformer) converts wave energy to produce electricity. The power is also used to run an electric bus which transports people around the community.

Teachers notes:
- Before starting the lesson, hold a class discussion about waves and wave energy to determine prior knowledge.
- Pupils read the text at the top of the page, then discuss the questions with a partner before completing the answers.

Publishing/Display ideas:
- Display pictures of ocean waves around the classroom.

Answers:
1. Wind blowing over the surface of the ocean.
2.
   - Waves hit buoy.
   - Water rushes in, compressing air in chamber.
   - Air causes turbine to spin a generator.
   - Generator makes electricity which is sent via cables to land.
3. Wave power stations should be situated in areas with strong winds such as off the west coast of USA, Europe and coasts of Japan and New Zealand.
4. *Advantages:*
   - renewable energy (means less reliance on fossil fuels)
   - need no fuel to run
   - produce no waste or pollution

   Disadvantages:
   - expensive
   - large and may interfere with shipping
   - reduce wave height and strength
   - need to be able to run (and stay intact) in severe storms but also with small waves

Additional activities:
- Research to find examples of where wave power is used in your country and elsewhere in the world. Mark the locations on a map of the world. Can any conclusions be drawn about the reasons why these countries are using these particular energy sources?
- Choose a technology which is being used (or tested) to harness wave energy to research further. For examples:
  - the 'Limpet', a wave power station operating in Scotland
  - the floating tube known as the 'Pelamis'
  - the 'CETO' which is moored to the sea floor (being trialled off the coast at Fremantle, Western Australia)

Alternative energy sources

Wave energy

Waves are caused by the wind blowing over the surface of the ocean. Ocean waves are powerful and generate large amounts of energy. Most of this wave energy is found in areas with very strong winds, such as the west coast of the Unites States and Europe, and the coasts of Japan and New Zealand.

Finding a way to capture wave energy so that it can be used to make electricity is a challenge. As a result, wave power stations are rare.

There are a number of different technologies being tested and used to harness wave energy.

One is a **buoy** with a large opening (a chamber) on one side. As waves hit the buoy, water rushes into the opening, compressing the air inside. A turbine is placed where the air is rushing in and out of the chamber. The air causes the turbine to spin a generator which makes electricity, which is sent via cables along the ocean floor back to land.

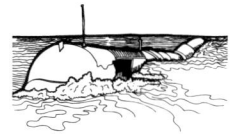

Another technology is a floating tube (known as a **'Pelamis'**) which bobs up and down on the waves causing its hinges to bend and pump hydraulic fluid to run generators to make electricity.

'Pelamis'

The equipment required to change wave energy into electricity is expensive, and these gigantic structures may interfere with shipping.

Further problems with harnessing wave energy involve the equipment required, as it:

– reduces the height and strength of the waves, which may have a negative impact on the coastlines

– must be strong enough to withstand rough weather and storms, but also be able to harness energy from small waves.

Wave energy is a renewable energy source. Once wave power stations are constructed, they need no fuel to run them and produce no waste or harmful pollution.

Let's hope the scientists of today and the future will design a more effective way to harness and convert this energy.

1. What causes waves? _____

2. List the steps of how a wave 'becomes' electricity using the 'buoy' technology.

- _____
- _____
- _____
- _____

3. Where should wave power stations be situated for best results? Give examples.

_____

4. List two advantages and two disadvantages of harnessing wave energy.

Advantages	Disadvantages
_____	_____
_____	_____

Alternative energy sources

Energy from waves – experiment

Objective:
- Conducts an experiment to explore wave energy.

Background information
- The 'Energy from waves' experiment shows how waves can produce moving air. On a large enough scale, moving air can be used to spin a turbine propeller or turbine to produce electricity.

Equipment/materials required:
Per group
- 2-litre juice container with lid
- straw
- Stanley knife™
- scissors
- large sink or tub
- water

Teachers notes:
- Teachers can use the Stanley knife™ to make a slit at the base of the juice containers, to help pupils cut away the bottom of the container with scissors.

Safety note:

Strict supervision is required for Steps 2 and 3 of the procedure. Depending on the class, teachers may like to prepare the lids with holes prior to the lesson.

Publishing/Display ideas:
- Take photos of the pupils conducting the experiment. Display them and ask pupils to add anecdotal notes about their observations. Pupils can use a word processor to publish their responses to the 'What you discovered' part of the experiment. Display these with the photographs.

Answers:
Teacher check

Additional activities:
- Research and create an information poster with diagrams, explaining how and why waves occur.
- Present a mini-debate between the owner of an energy company proposing to build gigantic wave power stations in the ocean, and the owner of a large shipping company, whose ships transport goods around the world.
- Write a narrative set in the future during a time when all the world's power is fuelled by wave energy. The title of the narrative is: 'The day the wind stopped!'

Alternative energy sources

Energy from waves – experiment

You are going to create a device which can produce energy from waves.

What you need:
- 2-litre juice container with lid
- scissors
- straw
- large sink or tub
- Stanley knife™ (for teachers)
- water

What you do:

1. Carefully cut the bottom from the juice container. (Ask an adult to help.)
2. The teacher will punch a hole in the lid of the container to fit the straw.
3. Fill the sink with water.
4. Push the straw through the lid and tightly secure the lid to the container.
5. Put the container in the water so the open bottom is just under the water.
6. Make waves.
7. Place your hand above the straw. What do you feel?

What happened:

Describe your observations and results.

_____
_____
_____

Draw a labelled diagram of your experiment.

Hints for drawing a diagram
- Use a sharp pencil. ☐
- Use a ruler for straight lines. ☐
- Use all the space you have. ☐
- Label all parts. ☐
- Rule a line from each piece of equipment to your label. ☐

What you discovered:

Explain how the experiment demonstrates how wave energy can be harnessed for power.

_____
_____
_____

Prim-Ed Publishing   www.prim-ed.com   Investigating renewable energy and sustainability

Alternative energy sources

Tidal energy

Objective:
- Reads texts and answers questions about tidal energy.

Background information:
- Tides are caused by the gravitational pull of the moon and sun, and also the rotation of the earth.
- In tidal power stations, the amount of electricity produced depends on the tidal range and the volume of water passing through the turbines.
- Further disadvantages of tidal energy include:
  – The construction of barrages can affect boating and shipping traffic and possible recreation developments in the area.
  – Sedimentation and movement of water in the area are affected.
- In 2008, a British tidal power company agreed to build a tidal power system in South Korea—larger than the Rance station in France.

Answers:
1. (a) Tidal range is the difference in water level between high and low tides.
   (b) Tidal barrage is a dam built across a river mouth or inlet to harness tidal energy.
   (c) Tidal turbine is a large turbine placed under the ocean offshore to harness energy from high tidal currents.
2. Tidal energy is renewable because tides will always continue to ebb and flow (go out and come in).
3. Choose from:
   – not many sites
   – expensive to build barrages
   – affects environment due to flooding of area and shoreline
   – marine animals have movement, mating and eating habits affected
   – shoreline vegetation reduced and altered
   – birds rely on mudflats for food
4. It is a renewable energy source so means fewer non-renewable fossil fuels (coal, gas, oil) are used. By decreasing the use of fossil fuels, emissions that harm our atmosphere are also reduced.

Teachers notes:
- Before starting the lesson, hold a class discussion about tides to determine prior knowledge.
- Pupils read the text at the top of the page, then discuss the questions with a partner before completing the answers.

Publishing/Display ideas:
- Source photographs of areas showing high and low tides and display around the classroom.
- Display photographs of tidal barrages, turbines and fences.

Additional activities:
- Research to find examples of where tidal energy is used (or could be used) in your country and elsewhere in the world. Mark the locations on a map of the world. Can any conclusions be drawn about the reasons why these countries are using these particular energy sources?
- Explain the difference between tides and waves, and include explanations of what causes both.
- Research and report on the environmental impact of the construction of the tidal power plant in the Rance estuary in France.
- Create a poster explaining how the moon affects the tides.

Investigating renewable energy and sustainability

Alternative energy sources

Tidal energy

The ocean is always moving. Twice a day, tides, caused by the gravitational pull of the moon and the sun, move large amounts of water. The difference in height of the water between high and low tides (the tidal range) varies across the world. In Canada, at the Bay of Fundy, the tidal range is 10.8 metres—the largest in the world!

Tidal barrages

Tides cause huge amounts of water to rush back and forth. When a large dam, called a tidal barrage, is built across the mouth of a river or inlet, big turbines built into the barrage walls are made to spin by the water rushing in and out. This energy is harnessed and can be used to drive generators which make electricity.

The largest tidal power station in the world, built in 1966, is in the Rance estuary in France.

Tidal energy:
– produces no greenhouse gases or other waste
– needs no fuel to run once it is built (it is free)
– is reliable as tides are predictable.

Unfortunately, building a tidal barrage is expensive and there are not many places around the world with a large enough tidal range to use tidal energy. Also, as construction of a barrage causes the area's tidal range to change and often flooding to occur, barrages have a big impact on the environment.

Environmental problems include:
– marine animals may have their movements restricted and mating and eating habits changed
– shoreline vegetation is reduced and changed
– birds rely on the tide uncovering mudflats so they can feed.

Tidal turbines

A new technology is being developed which sees rows of tidal turbines being installed under the ocean offshore where there are high tidal currents. The spinning turbines generate energy to create electricity.

There are many more sites available for tidal turbines, which have less impact on the environment.

1. Explain the following terms in your own words.

(a) *Tidal range* _____

(b) *Tidal barrage* _____

(c) *Tidal turbine* _____

2. Why is tidal energy a renewable energy source?
_____

3. List two disadvantages of using tidal power to create electricity.

• _____

• _____

4. Explain why it is important to us and to the earth for scientists to continue to design and test new technologies to harness tidal energy.
_____

Alternative energy sources

Tides and barrages

Objectives:

- Completes activities to investigate a tides table.
- Lists information required to determine if an area is suitable for the construction of a barrage.

Teachers notes:

- Pupils may require access to the Internet and reference books to complete Question 3 on the worksheet. Alternatively, pupils could work in small groups to complete this question.
- Note: The first piece of information the 'scientist' should collect about the area is its tidal range. It must be more than seven metres for a barrage to be considered.

Equipment/Materials required:

- Access to the library and Internet.

Publishing/Display ideas:

- Cut out the 'tide times' tables in the newspaper each day for one week and display them.

Answers:

1.

Place	Tidal range
Esperance	0.8 m
Albany	0.7 m
Bunbury	0.5 m
Fremantle	0.5 m
Barrack St	0.4 m
Geraldton	0.5 m

2. (a) Esperance
   (b) Barrack St.
   (c) 12:42 pm
   (d) 12:49 am
   (e) 13 hours 18 minutes
   (f) No, as the tidal range is too small.
3. Answers will vary

Additional activities:

- If you live near the coast, look in the local newspaper to find the height of high and low tides. What is the tidal range in the area?
- Study a weekly tide times chart. Do the times of the high and low tides change daily? (Yes) By how much? (52 minutes every 24 hours) Present an information poster about how the moon affects the tides.
- Paint scenes of a beach or river at high and low tides and display them side by side.

Alternative energy sources

Tides and barrages

1. Read the tides table and calculate the tidal range at each area.
(Tidal range = height at high tide – height at low tide)

Western Australia – Tide times					27 March
Place	Time	Tide	Time	Tide	Tidal range
Esperance	1.40 pm	1.1 m	6.01 am	0.3 m	
Albany	1.29 pm	1.1 m	5.26 am	0.4 m	
Bunbury	1.15 pm	1.0 m	3.11 am	0.5 m	
Fremantle	12.42 pm	1.1 m	2.00 am	0.6 m	
Barrack St	3.22 pm	1.0 m	5.23 am	0.6 m	
Geraldton	1.56 pm	0.9 m	12.49 am	0.4 m	

2. Answer the questions about the 'Tide times' table.

(a) Which place has the greatest tidal range? _____

(b) Which place has smallest tidal range? _____

(c) The first high tide occurs at what time? _____

(d) The first low tide occurs at what time? _____

(e) How much time has passed between high tide and low tide at Fremantle?

_____ hours _____ minutes

(f) For an area to be suitable to produce electricity from tidal energy, a tidal range greater than seven metres is required. Do you think a tidal power station should be constructed in the given towns in Western Australia?

(yes) (no) Explain your answer. _____

3. Imagine you are a scientist who has been asked to assess a particular estuary as to its suitability for the construction of a tidal barrage to harness energy to make electricity.
Write notes in the boxes below listing the information you would gather to make your assessment.

Geography of area	Marine animals	Vegetation	Shipping, boating and fishing

Alternative energy sources

Biomass energy

Objective:

- Reads texts and answers questions about biomass energy.

Background information:

- Plants absorb the sun's energy using a process called photosynthesis. When people and animals eat the plants, this (chemical) energy is passed on to the consumer, and is called bio-energy.

- In some developing countries, biofuels provide 90% of all energy, compared to only a small percentage in some countries of the industrialised world. For approximately half of the population of the world, wood or dung is the main source of energy.

- In the future, it may be possible for farms to use their waste to provide themselves and surrounding areas with electricity.

- Statistics say the average person throws away almost 2 kg of rubbish every day, (or 728 kg per year).

Answers:

1. Wood, rubbish, crops, alcohol fuels, landfill gas.
2. Ethanol is produced from fermenting crops of sugar cane, corn etc.
3. - Waste burned.
   - Converted into fuels such as methane gas, methanol and oils.
   - Methane burnt for heat to boil water.
   - Boiling water produces steam.
   - Steam turns turbines.
   - Turbines attached to generators to make electricity.
4. Biogas is methane gas which is released when biomass in landfills, sewage plants and on farms (manure) rots and decays.
5. *Advantages:*
   - renewable energy which can be replaced by growing more trees/crops
   - made on demand

   Disadvantages
   - produces air pollutants when burned
   - more is required to make electricity
6. No, because plastics, metals and glass are made from non-renewable energy and are not biomass, so they do not have stored solar energy within them.

Teachers notes:

- Before starting the lesson, hold a class discussion about biomass to determine prior knowledge.

- Read the text with the class. Ask the question:

  'Do we use biomass energy when we eat vegetables and fruit?' (Yes—when we eat vegetables and fruit, we are consuming the solar energy stored in them as they grew. Our bodies use this biomass energy to give us the energy to work and play.)

- Pupils read the text then discuss the questions with a partner before completing the answers.

Publishing/Display ideas:

- Source photographs and images of the main types of biomass—landfill sites, wood, rubbish, crops (corn, wheat, sugar cane) and display them.

Additional activities:

- Use the Internet to find the locations of countries that use biomass energy. Mark the locations on a map of the world. Can any conclusions be drawn about the reasons why these countries are using this particular energy source?

- Research to find out the special modifications required for vehicles to use a high percentage of ethanol in fuel. (Modifications are required to avoid damage to paint work and plastic and non-ferrous parts.)

Alternative energy sources

Biomass energy

Biomass is solar energy (energy from the sun) which is stored in plants and animals.

The five main types of biomass are: wood, rubbish, crops (corn, sugar cane etc.), alcohol fuels and landfill gas.

Wood

When wood is burned, the stored energy within it is released as heat which can be used to heat homes and in industry. The heat can also be used to boil water, which produces steam for making electricity.

Crops and alcohol fuels

When crops of corn and sugar cane ferment, an alcohol called 'ethanol' is produced, which can be used as fuel for transportation (joining with or replacing petrol).

Landfill rubbish

At landfills, much of the waste is biomass—such as food scraps, lawn clippings and newspapers.

When the waste is burned, a special conversion system can change the waste into fuels such as methane gas, methanol and oils. Methane can be burned to provide the heat for boiling water. Boiling water produces steam, which can be used to turn turbines that are attached to a generator to produce electricity.

Landfill gas

When biomass in landfills, sewage plants and on farms (manure) rots, it releases methane gas (biogas) which can be collected and used as a transportation fuel and to make electricity.

Unfortunately, burning biomass does pollute the air, though not as much as burning fossil fuels. Also, large amounts of biomass (compared to fossil fuels) are needed to create electricity.

Biomass is a renewable energy source as there will always be waste and we can always grow more trees and crops. Unlike other renewable energy sources, such as solar or wind, it can be generated on demand.

1. List the five main types of biomass.

- _____
- _____
- _____
- _____
- _____

2. How is the fuel 'ethanol' produced?

_____
_____

3. List the steps involved in using landfill rubbish to make electricity.

- _____
- _____
- _____
- _____
- _____
- _____

4. What is biogas?

_____
_____

5. Describe one advantage and one disadvantage of using biomass as an energy source.

Advantages

Disadvantages

6. Some rubbish in landfills can be burnt to produce electricity. Does this include plastics, metals and glass? (yes) (no)
On the back of this sheet, explain why.

Alternative energy sources

Energy from rubbish – experiment

Objective:
- Conducts an experiment to create a gas from decaying rubbish.

Background information:
- One area where biomass is found is landfill sites (rubbish dumps). Landfill gas is created when biomass waste, such as food scraps and lawn clippings, starts to rot and decompose in the ground. This gas will seep into the ground or into the atmosphere, unless it is captured. Landfill gas can be processed to create electricity!

Equipment/materials required:
Per group
- 30 dried beans or peas
- 3 zip-lock bags
- colander
- bowl
- water

Teachers notes:
- Note: The beans or peas need to be soaked overnight the day before the activity is carried out.
- Read the experiment with the class and ensure understanding before distributing equipment. Organise pupils into small groups to complete.

Publishing/Display ideas:
- Staple the bags to a display board in the classroom. Pupils use a word processing program to publish their observations and results, and display these with their set of bags.

Answers:
Teacher check.

Additional activities:
- Research to find out what it means to have a 'control' in an experiment. If there was to be a control bag of beans/peas in this experiment, where would it be placed? Discuss.
- Take digital photos of the bags during the week and on the final day. Display the results of the experiment as a photo diary with anecdotal notes.
- In a small group, discuss if you think the gas produced in the experiment could be used as a source of energy. Present your thoughts and ideas orally to another group or the class.

Alternative energy sources

Energy from rubbish – experiment

Can you produce a gas from decaying rubbish? Try the following activity below to find out!

What you need:

- 30 dried beans or peas
- 3 zip-lock bags
- colander
- bowl
- water

What to do:

1. Place the beans or peas in the bowl and cover them with water. Soak them overnight.
2. Use the colander to drain the beans.
3. Separate the beans into groups of 10. Place each group in to a zip-lock bag.
4. Squeeze out as much air as possible from the bag before it is sealed.
5. Place one bag in a warm, sunny place, another in a warm, shady place and the third in a dark place (such as a cupboard).

What you think will happen:

Which conditions do you think will cause the most gas to be produced? Order them from 1 (most gas) to 3 (least gas).

Bag	Rank	Bag	Rank	Bag	Rank
Warm, sunny place		Warm, shady place		Dark place	

What you observed:

Record your observations three times over a one-week period. Write in the number of the day you made your observations. Add sketches of interest.

Bag	Day ☐	Day ☐	Day 7 (final day)
Warm, sunny place			
Warm, shady place			
Dark place			

What you discovered:

Did the decaying beans produce a gas?    Were your predictions correct?

Which environment was best for producing a gas? _____

Investigating renewable energy and sustainability

Alternative energy sources

Geothermal energy

Objective:
- Reads texts and answers questions about geothermal energy.

Background information:
- The name 'geothermal' comes from two Greek words—'geo' meaning 'of the Earth' and 'thermal' meaning 'heat'.
- Harnessing geothermal energy can involve using the Earth's temperature near the surface, but it can also require drilling down many kilometres into the earth.
- Still today, the Maoris of New Zealand use hot rocks to cook food in the ground.
- Around the world, people swim in warm natural springs to help soothe body aches and pains.

Teachers notes:
- Before starting the lesson, hold a class discussion about geothermal energy to determine prior knowledge.
- Pupils read the text then discuss the questions with a partner before completing the answers.

Publishing/Display ideas:
- Find pictures or photographs of hot springs, geysers and volcanoes and display them.

Answers:
1. geysers, volcanoes, pools of bubbling mud, hot springs
2. reservoirs; steam; electricity
3.
   - Holes drilled into the Earth down to hot rocks.
   - Water pumped inside.
   - Water filters back up through cracks.
   - Water changed to steam.
   - Steam turns turbines to make electricity.
4.
   – renewable
   – reliable and available 24 hours per day
   – no polluting greenhouse gases emitted
   – uses small amounts of land
   – little impact on environment
5. – sites difficult to find
   – 'steam' may decrease or run out
   – may release harmful, toxic gases which are difficult to dispose of
6. New Zealand and Iceland.

Additional activities:
- Use the Internet to find the locations of countries that use geothermal energy. Mark the locations on a map of the world. Can any conclusions be drawn about the reasons why these countries are using this particular energy source? (For example, New Zealand uses geothermal energy as a result of underground volcanic activity.)
- The most active geothermal resources are usually found where volcanoes exist and earthquakes occur (along major plate boundaries). Most of the geothermal activity in the world occurs in the 'Ring of fire' surrounding the Pacific Ocean.

  Pupils conduct research to label a world map showing the location of 'The ring of fire'. They add to this map the areas where the best-known geothermal energy sources can be found.

Alternative energy sources

Geothermal energy

Geothermal energy is energy from the heat of the Earth. The centre of the Earth is extremely hot. A few kilometres below the Earth's surface (the crust), the temperature can be as much as 250 degrees Celsius. When this heat escapes to the surface, bubbling mud pools, hot springs, geysers and even volcanoes are seen.

There are two main ways this heat energy can be collected and used in geothermal power plants where it is used to spin turbines to make electricity.

Dry steam plants

In places such as Italy where there is steam in underground reservoirs, the steam can be piped directly to the surface into power plants to produce electricity.

Flash plants

Holes are drilled into the earth and water is pumped down to the hot rocks below. The water filters back up through cracks in the rocks and turns to steam when it reaches the surface. The steam turns turbines to make electricity. Most geothermal power plants are flash plants.

When the steam is cooled, it condenses back to water and is returned to the ground to be used again.

Scientists are concerned that some geothermal sites may 'run out of steam'. The energy from the Geysers in San Francisco, USA, powers a city of one million people; however, the steam being produced is slowly decreasing.

Geothermal energy sites can be difficult to find. Also, hazardous gases and minerals which are toxic and difficult to dispose of may come to the surface.

Geothermal energy is a reliable, renewable energy source, with energy available 24 hours a day. Harnessing this energy doesn't use very much land so there is not much impact on the environment, nor does it pollute the air with greenhouse gases.

New Zealand, the United States, Iceland and Japan all use this heat energy to run geothermal power stations.

Geysers geothermal power plant in Geysers, California, USA

---

1. What can be seen on the surface of the Earth when geothermal energy is escaping?

   _____ , _____ , _____ , _____

2. **Dry steam plants** are located above underground _____ where _____ is pumped to the surface and into the plants to make _____ .

3. List the steps involved in harnessing geothermal energy in a 'flash plant'.

   • _____   • _____

   • _____   • _____

   • _____

4. List two possible advantages of using geothermal power as an energy source.

   • _____   • _____

5. List two disadvantages of using geothermal power as an energy source.

   • _____   • _____

6. Research to find out which two countries that use geothermal power have land with volcanic activity.

   _____ and _____

Alternative energy sources

'Geothermal energy versus tourism' debate

Objective:
- Forms groups to carry out a debate about geothermal energy.

Teacher information:
- A debate is an organised argument between two teams. One team argues 'for' the topic. This is called the affirmative team. The other team, called the negative team, argues 'against' the topic.
- Before the debate begins, the *chairperson* states the topic and introduces the members of each team.

  Team members speak in the following order.

  (i) first member of the affirmative team

  (ii) first member of the negative team

  (iii) second member of the affirmative team

  (iv) second member of the negative team

  (v) third member of the affirmative team

  (vi) third member of the negative team

- The *timekeeper* allows each speaker a set time to speak, after which he/she rings a bell which tells the speaker that time is up.
- The *chairperson* deals with comments from the audience. He/She will also announce the winner, after consulting with the adjudicator.
- The *adjudicator* assesses the performance of both teams in relation to:
  – the content of the debate
  – how well the content is planned and organised
  – the way the speech is presented.

Answers:
Teacher check.

Publishing/Display ideas:
- Display photographs of popular geothermal tourist attraction sites (such as hot springs, geysers and volcanoes).

Equipment/materials required:
- The 'speakers' may need access to the library and the Internet to complete research to prepare their arguments.

Additional activities:
- Pupils evaluate how well they feel they carried out their jobs. They list the things they would change and improve if they were to take on the same roles again.
- Video the debate and play it back to the class. Pupils can volunteer positive criticisms to the nine people involved. After time has been allocated for the debate team members to polish their debate, they perform it to another class or the school.
- Research a geothermal energy tourist attraction (there are many popular sites in New Zealand and the Unites States). Pupils create a tourist brochure, pamphlet or poster to advertise the site.

Alternative energy sources

'Geothermal energy versus tourism' debate

There are many benefits to harnessing renewable geothermal energy to generate electricity. However, in some countries, such as New Zealand, there are concerns that using this energy will affect and reduce the power of the geothermal tourist attractions nearby, such as geysers and hot springs.

1. In groups you will be debating the topic:

> Harnessing geothermal energy to make electricity is more important than tourism to geothermal sites.

- The 'affirmative' team will argue that using geothermal energy for power is more important than tourism.
- The 'negative' team will argue that tourism and tourists' money are more important than energy.

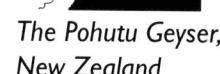

The Pohutu Geyser, New Zealand

There are nine people in a debate: a chairperson, three 'affirmative' team members, three 'negative' team members, an adjudicator (judge) and a timekeeper.

2. Form a group of nine and allocate responsibilities for the debate. Complete the boxes below to show who has been chosen for each job.

chairperson	timekeeper	adjudicator
affirmative speaker 1	affirmative speaker 2	affirmative speaker 3
negative speaker 1	negative speaker 2	negative speaker 3

3. Does each person understand his/her role? yes no

4. Complete the sentences below.

(a) My job is to be _____

(b) I have to _____

5. Use the space below to make notes or reminders about your job. You may also use this space to make notes about your arguments if you are a speaker.

6. Hold the debate. Which team was announced as the winner? Affirmative Negative
Do you agree? yes no

Alternative energy sources

Clueless crossword

Objective:
- Writes clues to match the answers of a crossword.

Teachers notes:
- Note: For best results, complete this activity once the comprehension activities on pages 41, 45, 49 and 53 have been completed.
- Collect crosswords from newspapers or crossword books for pupils to look at before they complete the activity on page 57. Pupils can familiarise themselves with how crossword clues are written.

Answers:
Teacher check.

Publishing/Display ideas:
- Publish and display the 'renewable energy' car bumper stickers from the 'additional activity' below. These can be printed onto paper with a self-adhesive backing for display.
- Display the 'renewable energy alphabet' poster from the 'additional activity' below.

Equipment/materials required:
- Pages 41, 45, 49 and 53 for reference.

Additional activities:
- Pupils work in small groups to brainstorm slogans that promote the use of one of the renewable energies listed. Pupils create colourful bumper stickers to illustrate their slogans. Display on classroom windows, bulletin boards and the back windows of cars.
- As a class, work together to create a 'renewable energy alphabet poster' that includes definitions and artwork for words related to renewable energy from 'A – Z'.
- Pupils use the words from the crossword to create a wordsearch for a friend to solve.

Clueless crossword

Alternative energy sources

Write the clues to match the answers to this crossword about four types of renewable energy sources.

biomass energy   geothermal energy   wave energy   tidal energy

ACROSS

1. _____
6. _____
7. _____
10. _____
12. _____
13. _____
15. _____
16. _____
17. _____

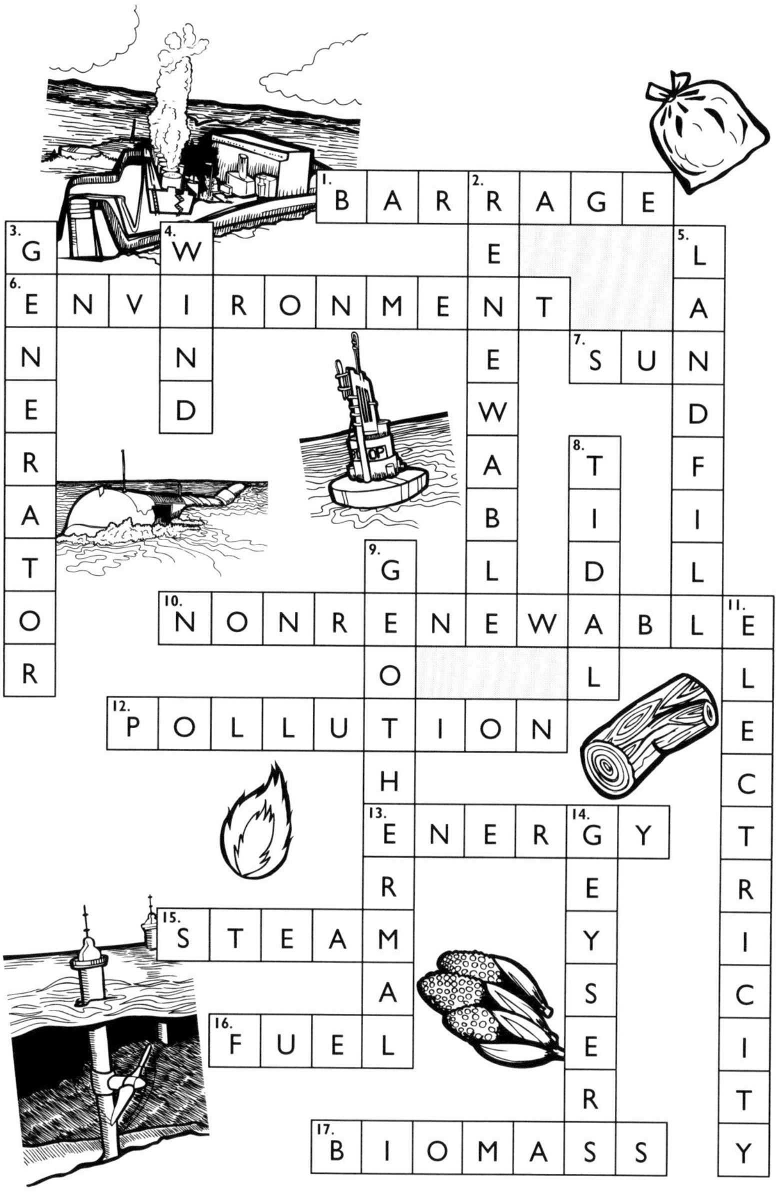

Across answers shown in grid: 1. BARRAGE, 6. ENVIRONMENT, 7. SUN, 10. NONRENEWABLE, 12. POLLUTION, 13. ENERGY, 15. STEAM, 16. FUEL, 17. BIOMASS

Down answers shown in grid: 2. GEYSER, 3. GENERATOR, 4. WIND, 5. LANDFILL, 8. TIDAL, 9. GEOTHERMAL, 11. ELECTRICITY, 14. GEYSERS

DOWN

2. _____
3. _____
4. _____
5. _____
8. _____
9. _____
11. _____
14. _____

Alternative energy sources

Advantages and disadvantages

Objectives:
- Lists the advantages and disadvantages of the four types of renewable energy.
- Assesses which renewable energy source is most suited to their local environment.

Teachers notes:
- Note: For best results, complete this activity once the comprehension activities on pages 41, 45, 49 and 53 have been completed.
- Pupils use their knowledge from previous activities plus information from the library/Internet to complete the advantages and disadvantages of each of the renewable energy sources listed.
- Organise the class into small groups to complete Question 2. Pupils can write the features of the local area on the back of the sheet or on a separate sheet of paper.

  Guide pupils by mentioning the following possible features:
  - geographical: coast and coastlines, rivers, presence of geothermal activity, wind speeds, tidal range etc.
  - businesses: timber and paper factories, farms, sewage plants, rubbish dumps, factories etc.

- Pupils assess and decide if their local area has features suitable for the creation and production of one or more of the renewable energy options.
- Allocate time for groups to share their ideas with the class.

Answers:
1. Refer to the following pages:
   - Page 41   Wave energy
   - Page 45   Tidal energy
   - Page 49   Biomass energy
   - Page 53   Geothermal energy
2. Answers will vary.

Publishing/Display ideas:
- Create and display four posters stating the advantages and disadvantages of each of the renewable energy sources. Include diagrams and artwork. Display in a public area of the school such as the library or hallway.

Equipment/materials required:
- Pages 41, 45, 49 and 53 for reference.

Additional activities:
- Once the worksheets have been completed, as a class, give each renewable energy source a star rating, from 0–5. An allocation of five stars means the pupils believe the renewable energy source should become a mainstream energy source in the future.
- If groups had different responses to Question 2, hold mini-debates so they can voice their reasons why they believe the renewable energy source chosen would be best suited to the local area.
- Write a letter to the editor of the most read newspaper in your area, expressing your concern about the harnessing of energy in the future. Include reasons why there should be greater development and use of renewable energy sources on a local and nationwide level.

Alternative energy sources

Advantages and disadvantages

1. Complete the table.

	Alternative energy	
	Advantages (Positive impacts of its use)	**Disadvantages** (Negative impacts of its use)
Wave energy		
Tidal energy		
Biomass energy		
Geothermal energy		

2. In your group, discuss features of your local area. List them on a separate sheet of paper.

(a) Would your local area be suitable for any of the above renewable energy options? ⬚ yes ⬚ no

(b) On the back of this sheet, explain your answer. If you answered 'yes', include which energy source.

Sustainable future

What is sustainability?

Objectives:
- Reads and answers questions about sustainability.
- Writes an acrostic poem about sustainability.

Background information:
- Over the past 200 years, humans have had a profound effect on the sustainability of those resources we need to survive. We have polluted the air by burning coal, gas and oil for energy and electricity; by using petrol and diesel as fuel for transportation; and we have consumed too many natural resources to make goods to meet our 'needs', some of which may never be able to be replaced.

- Acrostic poems do not have to rhyme, but each letter from the word 'sustainability' is used at the start of the lines of the poem. Each letter starts a new line.

Teachers notes:
- Read the text with the class. Ask pupils to highlight keywords and facts. Discuss these points before pupils complete the questions.

Publishing/Display ideas:
- Pupils publish their acrostics and display them.

Answers:
1. (a) True
   (b) False
   (c) False
   (d) True
   (e) False
   (f) False
   (g) True
2. Teacher check.

Additional activities:
- Pupils write and conduct a survey to determine the most popular mode of transportation taken to school by the pupils in the class (or whole school). Pupils present their findings as a graph. Pupils also discuss possible ways to motivate others to use environmentally-friendly modes of transportation such as walking and riding to school, where possible.

Sustainable future

What is sustainability?

Sustainable development is defined as 'meeting present needs while also taking into account the future', including the use of our environment and natural resources.

As people develop the land and use up the resources, we must ensure that what is done in the present does not have a negative outcome for future generations.

As far as we know, Earth is the only planet in the solar system which sustains life. Through pollution and overuse of natural resources, we are threatening that ability to sustain life. The state of the environment is directly influenced by our behaviours—whether we nurture it or mistreat it!

Natural resources come from the air, soil, water and living organisms. These are called 'renewable resources'. They include things such as wind and trees. 'Non-renewable resources' come from the subterranean areas of the Earth and include fossil fuels, such as coal, oil and gas.

People are in danger of using all the non-renewable resources, and are polluting many of the renewable resources. With the right attitude and determination we can build thriving cities where people live and work, while still maintaining resources for future generations.

We can all do our bit for sustainable development. Simple things such as using the car less, walking, cycling, using public transport, being more energy conscious and recycling waste can all make the environment a cleaner place for the future.

1. Use the information to answer the questions below.

(a) All species on Earth depend on each other. ☐ True ☐ False

(b) There are many planets in our solar system which sustain life. ☐ True ☐ False

(c) Overuse of resources has no effect on human life. ☐ True ☐ False

(d) Natural resources can be renewed. ☐ True ☐ False

(e) Coal is a renewable resource. ☐ True ☐ False

(f) It is impossible to have development and sustain resources at the same time. ☐ True ☐ False

(g) Little things we do, such as using public transport, can help the environment. ☐ True ☐ False

2. Use the word 'sustainability' to write an acrostic to show your understanding. Include a picture.

S _____
U _____
S _____
T _____
A _____
I _____
N _____
A _____
B _____
I _____
L _____
I _____
T _____
Y _____

Sustainable future

Energy-efficient homes

Objective:
- Reads and answers questions about energy-efficient homes.

Background information:
- Other aspects of energy-efficient home designs include using doors rather than an open-plan design to retain heated or cooled air; including solar hot water systems; the use of rain water tanks or adaptations to use 'grey' water for garden use etc.
- Many government departments assist home builders who want to build more energy-efficient homes. They can provide compass cards to help owners orientate their home best, supply brochures with information about energy-efficient heating, cooling and other appliances or brochures which give simple tips to save energy in the home.

Answers:
1. Orientation places the house on the block of land so that it utilises the sun's rays for heating in winter while excluding summer heat.
2. Answers may be chosen from the following list: size, placement, curtain or blind treatment, tinted glass, reflective film, double glazing.
3. (a) Insulation forms a barrier to keep heat in during winter and to reduce the amount of heat which enters the house in summer.
   (b) Building materials vary in their ability to keep heat in during winter and out during summer. Double brick is better than brick veneer but brick veneer is better than weatherboard, fibre cement and other lightweight materials.
4. Deciduous trees, planted correctly, will provide shade in summer but allow winter sun to warm the home. Grassed areas and large areas of paving or concrete shaded by trees help to reduce heat reflected into the house.

Teachers notes:
- Before starting the lesson, hold a class discussion about what it means to be 'energy-efficient' to determine prior knowledge.
- Pupils read the text then discuss the questions with a partner before completing the answers.

Publishing/Display ideas:
- Source and display pictures and photographs of energy-efficient, sustainable homes.

Additional activities:
- Draw a simple plan of the design of your home. Include compass directions; patios; courtyards; grassed, paved or concrete areas; arrows for hot and cold breezes and summer sun, large shade trees and shrubs.

  Give your home a rating to show how energy-efficient the design is (5 stars is the best).

- Investigate ways the school layout and design is energy-efficient and then list ways to improve the energy efficiency level.

- Visit a home display centre and collect brochures of energy-efficient buildings, products and appliances.

Sustainable future

Energy-efficient homes

A truly energy-efficient home begins with the orientation and design of the house itself. The most energy-efficient house is one which is placed on a block of land in such a way that it uses the sun's heat to warm the home in winter but excludes too much heat in summer. In the Northern Hemisphere, this means having windows and the most-used rooms, such as living rooms, facing south to receive maximum winter sun.

The size and placement of windows affects the amount of heat that enters a home. The type of curtains or blinds used on windows also makes a difference. Thick curtains can keep out or retain heat in rooms. Skylights in dark rooms reduce the need to turn on lights. Tinted glass, reflective film and double glazing on windows reduce the amount of heat leaving a room.

Insulation in the roof and walls of a home forms an efficient barrier to heat flow, reducing heat lost from a home in winter and the amount entering in summer.

The types of materials used to build a home can also make a big difference. Double brick walls are more efficient than brick veneer, while weatherboard, fibre cement and other lightweight walls are the least efficient as they heat up and cool down quickly. Concrete floors are more efficient than timber. Even painting the exterior of a home a lighter colour will help reflect summer heat.

Providing cross ventilation to capture cool breezes in summer and draught proofing around doors and windows to prevent heat loss in winter will help as well.

Plants and trees can also be used to make a home more energy efficient. Deciduous trees and vines planted correctly will provide shade in summer but allow winter sun to warm the home. Grassed areas and large areas of paving or concrete shaded by trees help to reduce heat reflected into the house.

A home designed for the local climate with energy efficiency in mind not only saves money on energy costs for its occupants, but may also reduce the impact that humans have on the environment.

1. What effect does orientation of a house have on energy efficiency?

_____

_____

2. List two ways windows can be energy-efficient.

- _____

- _____

3. Write an explanation to show how each affects energy efficiency.

(a) insulation: _____

_____

(b) building materials: _____

_____

4. On the back of this sheet, explain how plants and trees can be used to enhance energy efficiency.

Sustainable future
Designing an energy-efficient home

Objective:
- Plans and designs an energy-efficient home.

Background information:
- Builders and future home owners can take advantage of renewable energy sources, cutting down on their energy bills and helping to minimise the impact of energy use on the environment.

Tips for building energy-efficient homes:

NOTE: *Heating the home is the biggest part of an energy bill and insulation is one way to reduce these costs!*

- Overhangs (eaves) keep the summer sun out, while windows allow cooling breezes through.
- Positioning the house so the low winter sun can shine through windows to help heat the home.
- Installing a solar water heater.
- Positioning garages and closets on walls facing to the north, to maximise heat.
- Maximising carpet areas, instead of using tiles and concrete.
- Ensuring window placement allows ventilation.
- Installing insulation in the walls, floor and roof.
- Using energy-efficient whitegoods. (Checking the energy star rating—the more stars, the better.)
- Minimising electrical needs.
- Reflective roof with radiant barriers (made of reflective materials).
- Installing high-performance (double-glazed) windows.
- Installing solar-powered gate openers, outdoor lighting and water pumps.
- Sealing unwanted gaps and openings (especially near doors and windows).
- Sealing all electrical and plumbing fittings.
- Using fluorescent light bulbs.
- Installing ceiling fans.

Teachers notes:
- Pupils work in small groups and require adequate time to brainstorm their ideas about each part of the building design. They may also require time to research energy-saving ideas in the library and on the Internet.

Equipment/Materials required:
- Large sheet of craft paper, fine-tipped marker pens, access to the library and the Internet

Publishing/Display ideas:
- Display each group's design of its energy-efficient home.
- If models are built, use a digital camera to take photographs of the pupils building the models at different stages and display them with their house models and plans.

Answers:
Teacher check.

Additional activities:
- Display each of the designs. Ask pupils to view them and make constructive comments to the designers.
- Pupils build models of their energy-efficient homes.
- Conduct insulation experiments, such as using different materials to keep a can of drink cool or a jar of hot water warm.

Sustainable future

Designing an energy-efficient home

You and your team of builders and designers have been asked to submit a proposal for an energy-efficient home that incorporates renewable energy sources. Remember that heating is the biggest part of an energy bill and insulation is one way to reduce these costs!

1. Discuss each area of the building. Design and note your ideas, showing how you plan to construct and furnish the home so it is energy-efficient. Use the library and Internet to help you with energy-saving ideas.

Position of house (Which direction is it facing?)	Windows
Insulation	Heating
Floor coverings	Appliances
Roof/Shading	Ventilation and cooling

2. On a separate sheet of paper, draw detailed plans of your house. Include notes explaining how and why your building design is energy-efficient. Submit your proposal!

Sustainable future

Green vehicles

Objective:
- Reads texts and answers questions about 'green vehicles'.

Background information:
- People have used renewable energy sources for transportation for thousands of years—walking, riding animals and travelling in sail and rowing boats. Over a century ago, people began using non-renewable fossil fuels, which emit harmful pollution, for transportation.
- Motor vehicles emit hydrocarbons, nitrogen oxides and carbon monoxide, as well as large amounts of carbon dioxide, which may be trapping heat, causing the Earth's climate to change.

Answers:
1. A 'green vehicle' is a car that creates less damage to the environment when compared to traditional cars that run solely on petrol.
2. Due to traditional cars polluting the atmosphere.
3. Biomass—plants and animal waste.
4. **Electric vehicles:**
   What are they?
   – Vehicles run by battery pack.
   Positives
   – Emit no pollution when driven.
   Negatives
   – Batteries need recharging overnight (so not good for long-distance travel).
   – Also, batteries recharged by electricity which is generated by fossil fuels.
   Hybrid vehicles:
   What are they?
   – Vehicles run on petrol and electricity.
   Positives
   – Less petrol being used so less pollution.
   Negatives
   – Still using petrol to run, so still reliant on fossil fuels.

Teachers notes:
- Before starting the lesson, hold a class discussion about 'What is a green vehicle?' to determine prior knowledge.
- Pupils read the text at the top of the page, then discuss the questions with a partner before completing the answers.

Publishing/Display ideas:
- Source and display pictures and photographs of the green vehicles being sold in your country.

Additional activities:
- For one week pupils keep a record of each time they travel in a car and the approximate distance of the trip. They total the number of kilometres travelled and use the table to calculate the amount of carbon dioxide pollution emitted (multiply km by CO_2).

Vehicle size	CO_2 emissions
Small	150 g/km
Medium	200 g/km
Large	250 g/km
Large 4WD	300 g/km

- Research the top five most fuel efficient cars on the market today. What makes are they (Toyota/Ford etc.)? What technologies allow them to be so fuel efficient? Choose one vehicle and create an information poster about it.
- Conduct research to write a fact file about hydrogen vehicles. Include their advantages and disadvantages.

Sustainable future

Green vehicles

A 'green vehicle' is a car that creates less damage to the environment when compared to traditional cars that run solely on petrol.

There are almost 820 million cars in the world. For every litre of petrol used in a motor vehicle, 2.5 kilograms of carbon dioxide is released into the atmosphere. Carbon dioxide is a major greenhouse gas.

Some car manufacturers have looked at this growing problem and are developing more and more 'green vehicles', giving people wanting to buy a new vehicle more choice. 'Green' cars, scooters, buses, taxis, vans and trucks are being manufactured and sold today.

REVA – all-electric vehicle

Electric vehicles

Electric cars receive their energy to run from battery packs and produce no pollution while they are being driven. These batteries do run down and need to be plugged in to a powerpoint and recharged for many hours (usually overnight). Due to this limitation, electric cars are best used as a city car.

Hybrid vehicles

2008 Toyota Prius Hybrid

Some 'green' cars have been created which use less petrol. For example, the petrol-electric hybrid cars use petrol and electricity to run. Although the engine is run on petrol, the battery and motor are electric, meaning less pollution.

Popular hybrid vehicles include the Honda Insight and Toyota Prius. In New York, USA, there are about 400 hybrid taxis on the road.

Alternative-fuel vehicles

An alternative-fuel vehicle is a vehicle which runs using a fuel other than the fossil fuels—natural gas and oil (petrol).

When plant and animal waste decays, a gas is produced which can be used to make fuel to power cars. This fuel is called 'biofuel'. Cars that use biofuels instead of petrol reduce the amount of harmful pollution released to the atmosphere.

Other types of 'green vehicles' include hydrogen cars, solar-powered cars and cars being run using the alcohol 'ethanol', made from sugar cane crops.

1. What is a 'green vehicle'?
_____
_____

2. Why do you think people are choosing to buy 'green vehicles' over traditional petrol-run cars?
_____
_____

3. What is powering a car run on 'biofuel'?
_____

4. Complete the table.

Green vehicle	What are they?	Positives	Negatives
Electric vehicles			
Hybrid vehicles			

Investigating renewable energy and sustainability

Sustainable future

Green vehicles – play

Objective:
- Plans and writes a play about the sale of a green vehicle.

Teachers notes:
- For best results, complete this activity after the 'Green vehicles' comprehension activity on page 67.
- Pupils are likely to list colour, speed and cost as the main features the buyer is looking for in a green vehicle:

  Guide them to consider other features, such as:
  - its fuel efficiency (i.e. how far it can travel on a litre of fuel)
  - the cost of the fuel
  - how much air pollution is produced
  - which energy source(s) it uses for power
  - special features or limitations.

- Discuss with the class the characteristics of the 'seller', such as what it takes to be a successful salesperson (i.e. product knowledge, can be persuasive etc.). Also, discuss which characteristics of a green vehicle he/she will promote and which features the seller will downplay.
- Pupils write their plays.
- In pairs, pupils read and comment on each other's plays. Once checked, pupils rehearse and perform their plays together.

Answers:
Teacher check.

Equipment/Materials required:
- Page 67 worksheet, 'Green vehicles', for reference.
- Internet access for pupils wishing to research a particular green vehicle further.

Publishing/Display ideas:
- Once checked by a teacher, pupils work in pairs to rehearse and perform their plays to another group or the class. Plays could be filmed and watched by the class.

Additional activities:
- Ask a local car dealer who sells green vehicles to come and speak to the class.
- Pupils create a humorous cartoon about a family travelling in their new electric or solar-powered car. What humorous event can occur during the trip? (For example, they can't find a powerpoint to recharge the car's batteries or there is total cloud cover and no sunshine etc.)
- Investigate why solar-powered cars are considered by some to be not practical for day-to-day use.

Sustainable future

Green vehicles – play

You are going to plan and write a play about a person who visits a car dealer to purchase a 'green vehicle'.

1. Complete the information about the play.

Title _____

Characters

Buyer: _____

Seller: _____

Profile of the 'buyer'

- Occupation: _____
- Lives: City ☐   Country ☐
- Budget: _____
- Main uses of the vehicle.

  _____
  _____

Profile of the car dealer

- Name: _____
- Types of green vehicles available for purchase.

  _____

Setting

_____
_____

Features the buyer is looking for in a 'green vehicle'.

2. (a) Write the play. Remember to add 'directions' to describe each character's movements and how he/she is speaking (curiously, loudly etc.). Directions are written inside brackets. Continue on the back of this sheet.

(b) Work with a partner to read and comment on each other's plays. Rehearse and perform your plays.

Characters	Dialogue
_____	_____
_____	_____
_____	_____
_____	_____
_____	_____
_____	_____

Sustainable future

Sustainable transportation

Objective:

- Designs a futuristic transportation vehicle powered by a renewable energy source.

Teachers notes:

- Pupils work in small group to plan and design their sustainable modes of transportation. Check the designs and materials required before the pupils begin making their models as some materials may need to be ordered or found.

Answers:

Teacher check.

Equipment/Materials required:

- Pupils will need to supply the teacher with a list of materials required to construct their models (within reason). Some tools could be borrowed from the art/design classes for construction. Be cautious when pupils are using craft knives.

Publishing/Display ideas:

- Display each group's design.
- Take digital photographs of the pupils building their models and display them with each group's design.

Additional activities:

- Pupils present their design and model to the class, explaining how the vehicle works and why it is environmentally friendly. Pupils could also discuss the advantages and disadvantages of the design when compared to the transportation used today.
- Groups compare their design with the finished model. How do they compare? Pupils can record on a self-assessment sheet how they could improve their model, how their group members worked together and so on.

Sustainable transportation

Sustainable future

People have used renewable energy sources for transportation for thousands of years—walking, riding animals (such as horses, donkeys and camels) and travelling in boats. Less than 200 years ago, people started to use fossil fuels for transportation. These fuels are not renewable, so they will eventually run out. They also cause pollution.

1. Consider the renewable energy sources below. In your group, invent a future mode of transportation that is powered by one or more renewable energy source. Complete the plan for your vehicle below.

Design brief

Group members:

Design purpose:

Initial ideas:

Renewable energy used:
- solar energy ☐
- wind energy ☐
- hydropower ☐
- biomass ☐
- tidal energy ☐
- wave energy ☐
- geothermal energy ☐ _____ ☐

Design sketch (with labels showing materials used):

How does it work?

2. When you are happy with your design, transfer it onto a large sheet of art paper. Be accurate and neat.

3. Make a small-scale model of your vehicle and present it to the class.

Sustainable future

Energy use in your community

Teachers notes:

- Prior to this activity, contact some of the sources to check that they have the information needed and to find out how receptive they are to being contacted by school pupils.
- Pupils work in groups to research energy use in their community. If they are to contact organisations using the telephone, some lessons on telephone manners may be helpful.
- Pupils present their findings as an oral report. Show the class how to use note cards for public speaking. Main points to be discussed are written on the cards (not every word).

Answers:

Teacher check.

Equipment/Materials required:

- Internet access, telephone access (and telephone books), note cards for oral reports.

Publishing/Display ideas:

- Pupils present their findings at an assembly.
- Pupils edit and publish their finished speeches using a word processor. Collate the information to make a class book about energy use in the community.

Additional activities:

- If a video camera is available, film the pupils presenting their oral reports. Play the footage back to the class and discuss effective public speaking techniques.
- Ask pupils to write thankyou letters to any people or organisations who helped them with their research.

Investigating renewable energy and sustainability

Sustainable future

Energy use in your community

Which energy sources power your community? Are there renewable energy sources being used for power in your local area? Is your community aware of the importance of being energy conscious and of the impact of using fossil fuels on the environment?

1. Investigate which energy sources are being used in your community and if any renewable energy sources are being used or tested. You will need to use the Internet and telephone to help you locate this information. Some companies may send you special packs of information.

Try these sources:

- *Local council (website/telephone/visit)*
- *Building boards*
- *Energy boards (water/electricity/gas etc.)*
- *Environmental organisations*

2. Make notes of your findings below. Be accurate with facts, vocabulary and statistics!

Keep a record of your information sources:

3. Present your findings as a brief oral report to your class. Include:

- facts and data concerning energy in your community
- your opinion on energy consumption in your community
- suggestions on how your community could be more energy aware
- your sources of information.

Investigating renewable energy and sustainability